# Surface NMR for Hydrogeology

A user's guide

# Surface NMR for Hydrogeology

A user's guide

**Anatoly Legchenko**

*IRD, Université Grenoble Alpes–IGE, Grenoble, France*

**IOP** Publishing, Bristol, UK

ISBN 978-0-7503-3155-5 (ebook)
ISBN 978-0-7503-3153-1 (print)
ISBN 978-0-7503-3156-2 (myPrint)
ISBN 978-0-7503-3154-8 (mobi)

DOI 10.1088/978-0-7503-3155-5

Version: 20211101

IOP ebooks

British Library Cataloguing-in-Publication Data: A catalogue record for this book is available from the British Library.

Published by IOP Publishing, wholly owned by The Institute of Physics, London

IOP Publishing, Temple Circus, Temple Way, Bristol, BS1 6HG, UK

US Office: IOP Publishing, Inc., 190 North Independence Mall West, Suite 601, Philadelphia, PA 19106, USA

Supplementary material is available for this book from http://iopscience.iop.org/book/978-0-7503-3155-5.

# Contents

# Preface

For human beings, water is the most important resource. Groundwater is the principal source of drinking water in many regions of the world. Sustainable groundwater management requires knowledge of the extent and capacities of groundwater reservoirs. Different aspects relating to groundwater resources are studied in the branches of Earth science: geology, hydrogeology and applied geophysics.

Since long ago, surface geophysical methods have been used for groundwater prospecting. Recent advances in electronics and computer sciences boosted the development of geophysical instruments and technologies. Improved performance of existing geophysical tools and the development of new geophysical methods extend the application of geophysical prospecting beyond the delineation of geological formations, and give way to more extensive use of geophysics for characterizing groundwater reservoirs and in environmental science. One of the new geophysical methods is magnetic resonance sounding (MRS), also known as surface nuclear magnetic resonance (SNMR). This method is selectively sensitive to groundwater and was developed for hydrogeological purposes.

Technological development not only boosted the performance of geophysical tools but also their complexity and cost. Often, professionals in geophysics and hydrogeology face the problem of choice, because they cannot afford an expensive instrument with unknown parameters and prefer to use a known one. It may happen even when this other instrument or method would be better adapted to their study. Thus, an active learning of advanced geophysical tools using the numerical modeling approach may help to render professional activities more effective in term of the quality/price ratio.

This book, written based on the author's teaching experience, is a textbook for students of different backgrounds and professionals in hydrogeology and geophysics. Such a broad audience imposes a step-by-step description of the MRS method, starting with a general presentation of the geophysics, hydrogeophysics and hydrogeology. If not interested, readers may skip the boring pages.

Existing software packages do all the mathematical work for any geophysical or hydrogeological method including MRS. It gives rise to a wrong belief that the user does not need to understand what and how the computer does the job and it is sufficient just to learn how to run the modeling software. Because of this wrong belief, two chapters present the minimum necessary background material. The MRS forward modeling procedure presented in chapter 4 explains the forward modeling and shows how to use this procedure. A few basic examples animate this presentation. The MRS inversion presented in chapters 5 and 6 is focused on the uncertainty estimate. This section aims to teach not how to do inversion but how to guide computers to do inversion correctly, how to understand inversion results and how to use these results in practice. The last chapter gives a few field examples showing how to apply the knowledge gained when reading the content presented in the previous chapters.

For training, this book assumes the use of two software packages. The SAMOVAR software included in the book is developed by the author for modeling and inversion of MRS measurements. The HYDRUS-1D software developed by the PC-Progress for the unsaturated flow modeling is the free one-dimensional version of the more extensive HYDRUS package (https://www.pc-progress.com). SAMOVAR allows easy data exchange with HYDRUS. Both SAMOVAR and HYDRUS-1D are the very basic programs with the minimum options to select, which helps to learn faster. More advanced versions of these software packages are commercially available. However, the presented modeling procedure can be done with any other programs developed for processing MRS data and for water flow modeling.

Thus, with this book, students can learn about MRS and practice using this method. Professionals can make an evaluation of the method performance before using it in practice.

# Author biography

## Anatoly Legchenko

Anatoly Legchenko graduated from the Electrotechnical Institute of Novosibirsk (USSR), as an engineer in computer sciences and electronics. His PhD diploma in physics and mathematics was awarded by the Computing Center of the Siberian Branch of the Russian Academy of Sciences after presentation of the thesis 'Inverse problems in geophysical NMR tomography'. In 2001, the diploma Habilitation à Diriger les Recherches in geophysics was awarded by the University Paris XI (France).

Between 1981 and 1995, Anatoly was employed as an engineer and then as a research scientist in the Institute of Chemical Kinetics and Combustion of the Siberian Branch of Russian Academy of Sciences (ICKC, Novosibirsk, Russia). In 1995, he joined the Bureau de Recherches Géologiques et Minières (BRGM, France), and in 2003 the Institute de Recherche pour le Développement (IRD, France).

Currently, Anatoly is a senior scientist in IRD attached to the Université Grenoble Alpes (UGA, France).

**IOP** Publishing



# Chapter 1

# Introduction

## 1.1 Geophysics

Geophysics is a branch of Earth science that deals with the rocks and the physical processes occurring on, in, and above the Earth (Lowrie and Fichtner 2020). Applied geophysics is a branch of geophysics, which uses physical methods to measure the physical properties of the subsurface. The variety of geophysical activities explains a wide range of investigation methods (Knödel *et al* 2007). Figure 1.1 shows a broad classification of geophysical methods.

Depending on the acquisition approach, geophysical methods may be active or passive. Active methods require an artificial signal to be generated (for example, a current in the loop). Passive methods measure naturally occurring fields or properties of the Earth. Both active and passive methods can be airborne or space-based, applied in boreholes from the surface. Rock and soil samples can be studied in a laboratory. Geophysical methods are specialized and focused on measuring different physical properties. For example, seismic methods measure the velocity of seismic waves in rocks and electrical and electromagnetic methods measure the electrical resistivity.

In the great majority of applications, geophysical methods measure an integral signal corresponding to all the formations together. Deconvolution of this integral signal allows recovering physical properties of each individual formation. In geophysics, this procedure is called inversion.

Inversion of geophysical measurements requires the defining of a geophysical method and the corresponding measuring setup (figure 1.2). Depending on the method, the subsurface is represented by formations with physical properties the applied method will measure. For example, if we apply the electrical resistivity method, then we define an array of electrodes and represent the subsurface by formations with different electrical resistivity. The mathematical model allows computing a data set corresponding to this model of the subsurface. Inversion comprises building a model of the subsurface so that the synthetic signals are equal

doi:10.1088/978-0-7503-3155-5ch1

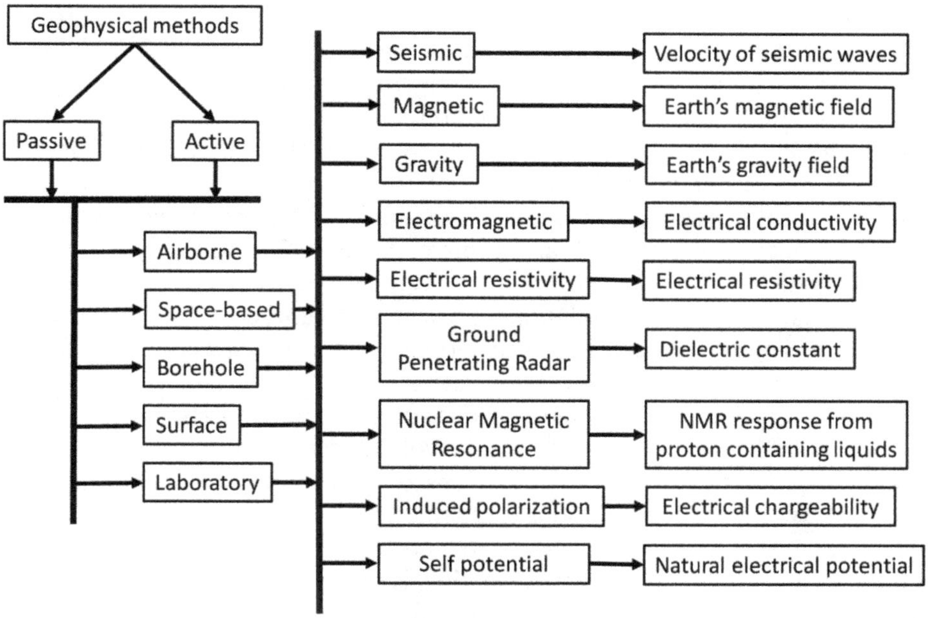


**Figure 1.1.** Broad classification of geophysical methods.

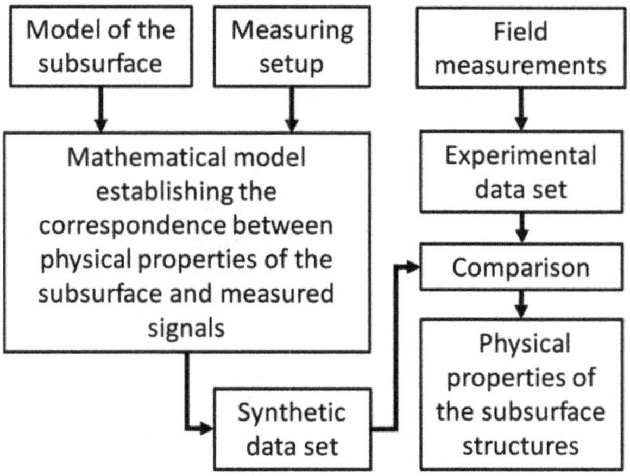


**Figure 1.2.** Deconvolution (inversion) of geophysical measurements.

to the measured signals. The solution of a geophysical inverse problem is not unique, which gives rise to the equivalence problem. It means that many models of the subsurface may produce very similar signals. The equivalent models form a solution space. To select the best model out of the solution space is a special procedure that requires the use of more information about the solution or assumptions made regarding the solution shape. A model of the subsurface provided by inversion is called the inverse model.

An inverse model shows subsurface structures with different physical properties identified by geophysical measurements. But geologists and hydrogeologists are looking for geological, geotechnical and hydraulic properties of the subsurface. Geophysics cannot provide this information without using empirical correlations between the physical properties and other properties that characterize subsurface structures, as shown in figure 1.3. Some established correlations are very robust and allow easy identification of geological formations. For example, crystalline rocks have a high electrical resistivity and a high velocity of seismic waves. Metal ore bodies and saltwater aquifers have a low electrical resistivity (Telford *et al* 1990). However, many reported correlations are less pronounced and are valid only in areas where they have been developed. Straightforward expansion of these correlations to other regions may give inaccurate results.

Geological formations are heterogeneous and physical properties measured using geophysical methods are averaged over the volume defined by the measuring setup. These bulk properties include contributions of all components that make up the subsurface. When using correlations established for each component, it is necessary to make deconvolution of the measured bulk properties. For example, the bulk electrical resistivity of an aquifer formation is a nonlinear function of the water content, the water salinity and the clay content. Deconvolution allows for estimating the contribution to the bulk resistivity of each of these components. Such a deconvolution is often a hard task and requires the use of all available data about the subsurface.

Figure 1.4 summarizes the principal steps necessary for interpreting geophysical measurements. First, inversion of measured data shows subsurface formations and quantifies their size, location and bulk physical property. At this step, geophysics provides structural information about the subsurface. For example, electrical resistivity measurements show variations of the electrical resistivity in the subsurface. Then, for each formation, the bulk physical properties should be converted into geological properties or other properties of the subsurface. If the measured physical property depends on only one component of the subsurface, then correlations can be

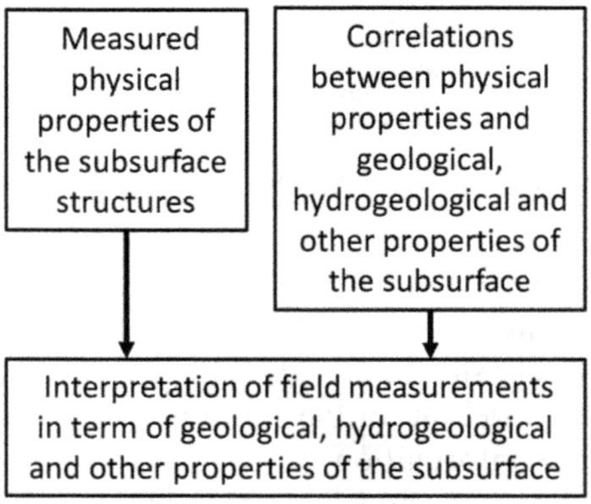


**Figure 1.3.** Characterization of a geological formation identified by the geophysical survey.

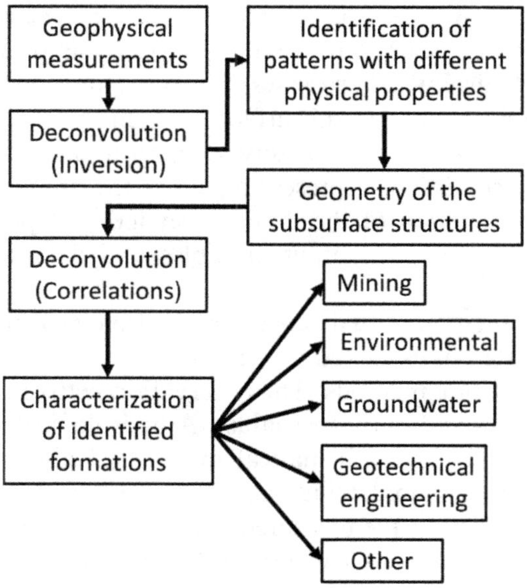


**Figure 1.4.** Interpretation of geophysical measurements.

straightforwardly applied. Otherwise, deconvolution should be applied. Finally, interpretation using calibrations completes the geophysical survey.

For demonstration, let us imagine an example of a geophysical study using electrical resistivity measurements. This study aims to estimate the porosity of an aquifer formation. The measured bulk resistivity of the water-saturated formation depends on the resistivity of water and the resistivity of the formation material. Porosity of this formation controls the contributions of water and of the material into the measured resistivity. If the resistivity of water is known, then the porosity can be estimated using Archie's law (Kirsch 2009). Archie's law is an empirical relationship developed for pure sand, and the porosity of a sand aquifer can be calculated straightforwardly using only the resistivity of water. But, when clay is present in the subsurface, correlations between the bulk resistivity and the porosity is a nonlinear function of the clay electrical resistivity, the clay content and the clay distribution throughout the aquifer formation (homogeneous, lenses, random patterns, etc). Deconvolution of the bulk resistivity into the clay contribution and the water contribution becomes much more complex.

## 1.2 Hydrogeology

Geology is an Earth science studying the rocks and the processes by which they change. Hydrogeology is a branch of geology that deals with the distribution and movement of groundwater in soils and rocks (De Marsily 1986). The prediction of future behavior of an aquifer system based on analysis of past and present observations is one of the principal tasks of a hydrogeological survey. For that, we have to solve the groundwater flow equation by approximating the real system by a hydrological model. A hydrological model can be developed using different approaches. In this book we use a modeling approach based on our understanding of the physics of the hydrological processes and corresponding equations to describe these processes. We represent the subsurface by domains with constant hydraulic properties (hydraulic conductivity, porosity, etc), set initial conditions at time zero and define boundary conditions representing either the physical boundaries of the domain, or an approximation of the domain beyond that point. Then, we apply hydrological and meteorological input variables and calculate output variables (Rushton 2003).

Figure 1.5 shows the conceptual scheme of hydrological modeling. The investigated water flow system is equipped by an observation network using observation wells and various recording methods. The observation network comprises measurements of input variables and output variables. A hydrological model approximates the water flow system by a mathematical model and allows computing output variables, considering measured input variables. This model is acceptable if the computed output variables fit the measured ones with the prescribed accuracy. For example, computed and measured water level variations versus time show equal (almost) values.

In porous media, some physical properties are measured with geophysical methods, for example, the electrical resistivity and the seismic velocity are linked with the quantity of groundwater in the formation. But a flow of groundwater may have a minor effect on variations of these properties and hence, geophysics cannot replace hydrological observations.

Figure 1.6 presents contributions of surface geophysical methods to hydrological modeling. The data set for modeling comprises input variables and structural information about the subsurface. Hydrological and meteorological observations represent the principal contribution to the input variables. Because of limited resolution, surface geophysical methods measure the geometry of a geological

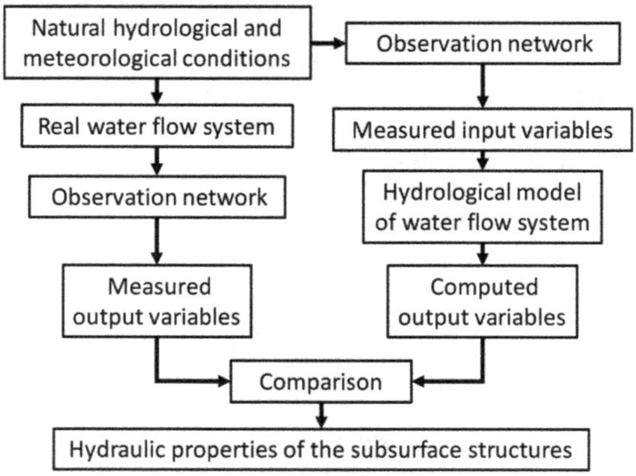


**Figure 1.5.** Conceptual scheme of the hydrological modeling framework.

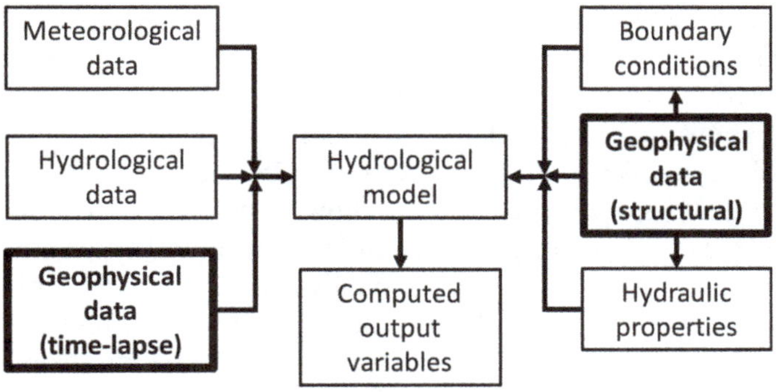


**Figure 1.6.** Contribution of geophysical methods to water flow modeling.

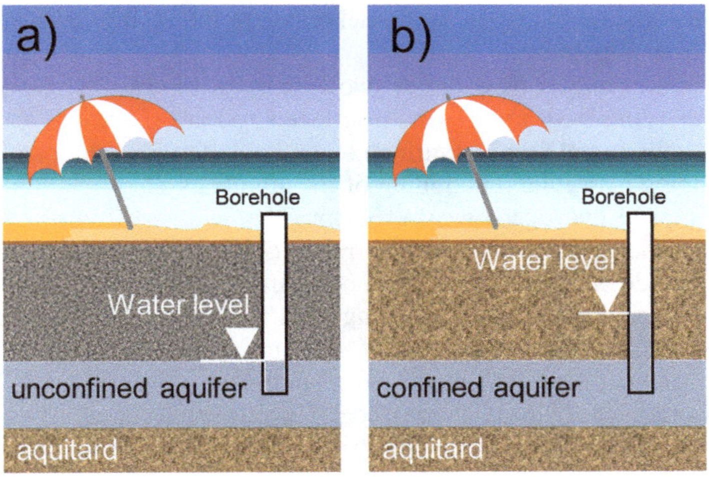


**Figure 1.7.** Correspondence between the hydraulic head and the top of water-bearing formation detected with geophysical measurements: (a) unconfined aquifer; (b) confined aquifer.

layout with an uncertainty no better than a few meters, and the top of water-saturated formations given by geophysical measurements may not correspond to the water level measured in a borehole. In an unconfined aquifer, a borehole measures the water level corresponding to the hydraulic head, and geophysics shows the top of the first water-saturated formation, which coincides with the hydraulic head (figure 1.7(a)). In a confined aquifer, the hydraulic head and the depth to the top of the first aquifer measured with geophysical methods may be not equal (figure 1.7(b)). Thus, geophysics cannot replace measurements in a borehole. Both the porosity and the hydraulic conductivity can be estimated only in the water-saturated part of the subsurface.

An insufficient accuracy of geophysical results does not allow for measuring minor changes in the physical properties related to groundwater flow and/or variations of the water level. It limits the application of geophysics to measuring input variables. However, when changes in the physical properties are large and can be observed with geophysics, these measurements provide useful data for hydrological modeling. For example, saltwater intrusion into the aquifer formation may cause large variations of the electrical resistivity. Note that one geophysical measurement shows a screenshot of the subsurface at a time of measuring and cannot quantify dynamics of groundwater flow. Only time-lapse measurements give access to quantification of groundwater flow. The principal contribution of a geophysical survey to hydrological modeling is structural information about the subsurface. It allows for selecting the groundwater flow model and the boundary conditions necessary for solving the groundwater flow equation. Under favorable conditions, geophysical measurements also allow for estimating the porosity and the hydraulic conductivity of water-saturated formations. When a hydrological model is developed, geophysical results help to decrease uncertainties in modeling results by constraining variations of the model parameters. Parameters justified by geophysical measurements make a hydrological model closer to the reality, thus increasing the model predictive capability.

For demonstration, let us consider two examples. Figure 1.8 shows an area of unknown extensions with unknown subsurface structure (Example 1). The project comprises installation of an observation network and development of a hydrological model describing hydraulic functioning of the aquifers system.

This is a typical hydrogeological problem. The fieldwork comprises continuous observation of the input and output flow rates ($Q_{\text{in}}$ and $Q_{\text{out}}$), drilling boreholes for measuring variations of the water level, performing a borehole logging, analyzing rock samples, and carrying out pumping tests. The boreholes number and their location depend on the complexity of the subsurface and the budget. If the subsurface structure is unknown, then one may use a uniform grid. Then, a few years monitoring of hydraulic parameters produces a data set sufficient for developing a hydrological model. The completeness of the model depends on the

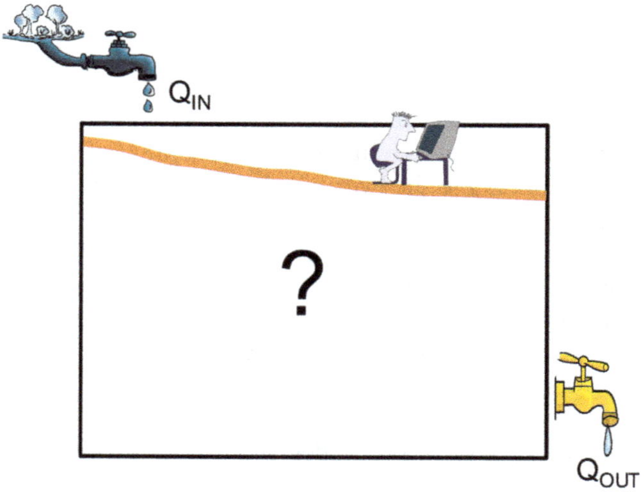


**Figure 1.8.** Example 1: data available for water flow modeling without geophysics.

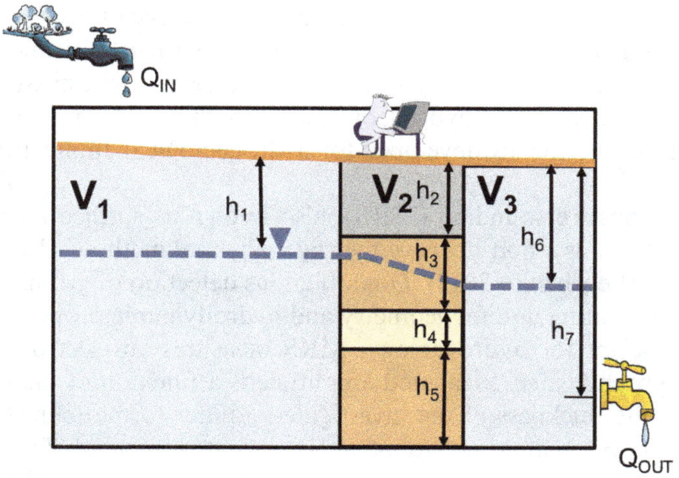


**Figure 1.9.** Example 2: data available for water flow modeling with geophysics.

modeling objectives and the data set. Different models may explain hydrological observations equally well. These models are the equivalent models. Choice of the best model depends on the observation duration, data quality, and amount of complementary information about the investigated aquifer.

The second project (Example 2) started by performing a geophysical survey. Figure 1.9 shows the subsurface structure derived from geophysical measurements.

The subsurface is composed of three distinct areas. A heterogeneous structure $V_2$ comprises four layers and separates two uniform formations of unequal size ($V_1$ and $V_3$). Geophysical measurements show the top of the aquifer. If the aquifer is unconfined, then the top of the water-saturated part in these three formations shows the hydraulic head. From $V_1$ to $V_3$, the water level varies from $h_1$ to $h_6$, suggesting that the transmissivity of the formation $V_2$ is smaller than that of $V_1$ and $V_3$. Both the porosity and the hydraulic conductivity can be estimated in the water-saturated part of the subsurface. Thus, geophysical results allow optimizing the implantation of boreholes and the design of the observation network. Time-lapse geophysical measurements show variations of the water content in the unsaturated zone, thus contributing to quantification of the aquifer recharge. Monitoring of this aquifer system will produce a data set for the hydrological modeling. For this project, geophysical measurements help to optimize the observation network and to constrain hydrological modeling, thus reducing the number of equivalent models.

## 1.3 Hydrogeophysics

Hydrogeophysics is a branch of applied geophysics that uses geophysical methods to estimate parameters and monitor processes important to hydrological investigations (Rubin and Hubbard 2006). Almost any geophysical method can contribute to groundwater investigations, but some of them are more adapted for this task. Choice of the survey method depends on geological setting, measuring conditions, project objectives, facility of the use, and the cost effectiveness of application. The most common are electrical and electromagnetic methods because of a link between groundwater and the electrical resistivity of the subsurface. Seismic methods and ground penetrating radar, widely used for engineering investigations, are also essential components of the hydrogeophysical toolbox. The magnetic resonance sounding method specifically developed for hydrogeological applications occupies a special place.

Magnetic resonance sounding (MRS), also known as surface nuclear magnetic resonance (SNMR), is a non-invasive hydrogeophysical method selectively sensitive to groundwater (Legchenko 2013). Unambiguous detection of groundwater and the possibility of estimating aquifer geometry and hydrodynamic properties make MRS particularly efficient for hydrogeology. MRS measures an electromagnetic signal generated by groundwater. Measured amplitude is a function of the water content, the depth and the thickness of the investigated aquifer formations. The relaxation time depends on the mean grain size of water-saturated material. The water content and the relaxation time allow estimating of the hydraulic conductivity of aquifers (Dunn *et al* 2002).

Detection of groundwater both in aquifers and in the unsaturated zone is the most reliable information delivered by MRS. In permeable soils and rocks with a saturated hydraulic conductivity over $10^{-5}$ m $s^{-1}$, the threshold of water detection depends on the measuring conditions and varies between 0.02 and 0.05. Water in low-permeable rocks with a saturated hydraulic conductivity less than $10^{-6}$ m $s^{-1}$ (clay, for example) is undetectable with MRS. MRS shows these rocks, even when saturated, as a dry material. If MRS detects groundwater, then an inversion of MRS measurements shows the geometry of water-bearing formations and the water content in each of them. Similar to other surface geophysical methods, MRS inversion is ill-posed and has multiple solutions. A limited resolution of MRS inversion does not allow for the resolving of small patterns and MRS yields results averaged over the area covered by the measuring loop. In aquifers, MRS measurements allow for the estimating of the basic parameters that characterize the aquifer productivity. The water content in a saturated porous media is close to the effective porosity and the hydraulic conductivity can be predicted using an empirical relationship that links up the hydraulic conductivity with the water content and the relaxation time being both a product of MRS inversion. Note that while MRS inversion straightforwardly gives the water content, an estimate of the hydraulic conductivity requires calibration.

Let us compare MRS with other popular geophysical methods. Figure 1.10 shows typical procedures of interpretation in terms of groundwater considering seismic, electrical/electromagnetic and MRS measurements.

Inversion of seismic measurements shows the location of geological formations with different velocities of seismic waves and quantifies the velocity in each identified formation. Then, the velocity in these formations has to be converted into the water content. The relation between the seismic velocity and the hydraulic conductivity is complex, resulting in poor characterization of hydraulic properties from seismic data. Electrical and electromagnetic measurements show formations with different electrical resistivity. Estimations of the water content using the resistivity is easy for

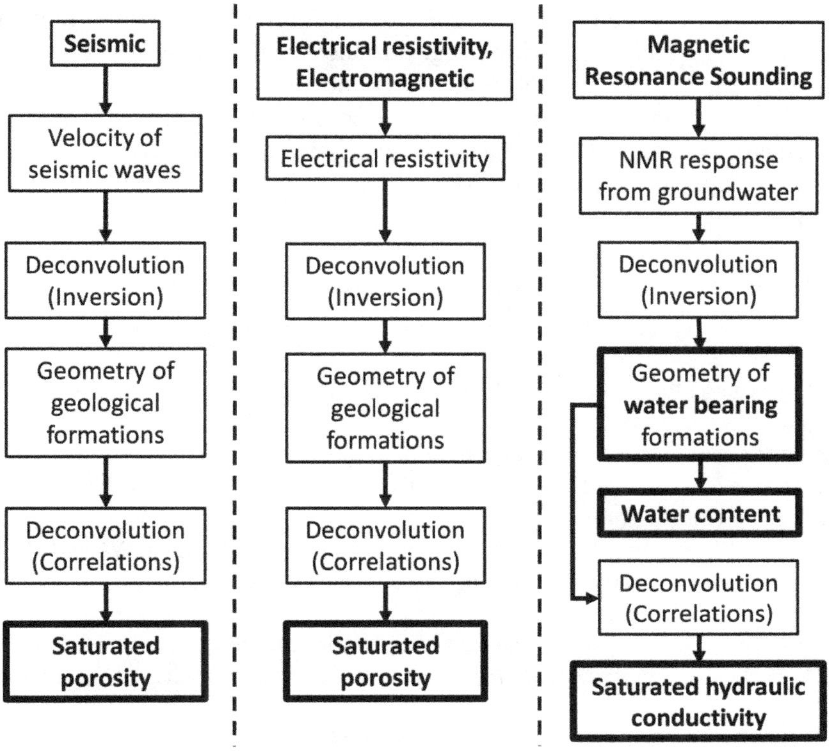


**Figure 1.10.** Typical interpretation procedure of three hydrogeophysical methods: seismic, electrical/electromagnetics and MRS.

pure sand and becomes much more uncertain when clay is present in the formation. Conversion of the resistivity into the hydraulic conductivity is challenging and, when possible, results are scale and site dependent. MRS inversion shows a vertical profile of the water content. Calibration can improve accuracy, but it is not strictly necessary. In a saturated porous media, the qualitative estimate of the hydraulic conductivity based on the relationship between the relaxation time and the mean size of pores is straightforward. A more accurate estimate based on site dependent empirical relationships requires calibration. Usually, MRS allows for the estimating of the saturated hydraulic conductivity with smaller uncertainty relative to other surface geophysical methods. Thus, MRS inversion shows the geometry of water-bearing formations, and inversion of seismic and electrical/electromagnetic measurements shows geological structures. Physical properties of a geological formation depend on various factors, and groundwater is only one of them. For example, the electrical resistivity of soils and rocks depends on the porosity, the water content, the water salinity, the mineralogy, the clay content and other factors.

Developed in Russia in the early eighties, MRS is now an established hydrogeophysical method. MRS instruments are fabricated in France, USA, Germany and China. They are commercially available, and the most popular are NUMIS$^{poly}$ of IRIS Instruments (France) and GMR$^{TM}$ of Vista Clara (USA). A few teams produce software for data processing and interpretation of MRS measurements. The cross verification of the computing programs shows that they produce very similar results. Different numerical implementations and inversion procedures explain the observed insignificant discrepancy. So, we recommend to users any of the established software packages. In this book, we use the tutorial version of the Samovar software developed for processing MRS data recorded with the NUMIS family instruments.

## Bibliography


De Marsily G 1986 *Quantitative Hydrogeology: Groundwater Hydrology for Engineers* (New York: Academic)

Dunn K J, Bergman D J and Latorraca G A 2002 *Nuclear Magnetic Resonance Petrophysical and Logging Applications* (Oxford: Elsevier)

Kirsch R 2009 *Groundwater Geophysics* (Berlin: Springer)

Knödel K, Lange G and Voigt H J 2007 *Environmental Geology: Handbook of Field Methods and Case Studies* (Berlin: Springer)

Legchenko A 2013 *Magnetic Resonance Imaging for Groundwater* (London: Wiley-ISTE)

Lowrie W and Fichtner A 2020 *Fundamentals of Geophysics* (Cambridge: Cambridge University Press)

Rubin Y and Hubbard S S 2006 *Hydrogeophysics* (Dordrecht: Springer)

Rushton K R 2003 *Groundwater Hydrology: Conceptual and Computational Models* (New York: Wiley)

Telford W M, Geldart L P and Sheriff R E 1990 *Applied Geophysics* (Cambridge: Cambridge University Press)

**IOP** Publishing

Surface NMR for Hydrogeology
A user's guide
**Anatoly Legchenko**

# Chapter 2

# Magnetic resonance sounding

## 2.1 Background

The magnetic resonance sounding (MRS) method is a geophysical application of the nuclear magnetic resonance (NMR) phenomenon with a theoretically complex theory behind the technique. Subatomic particles (electrons, protons and neutrons) can be imagined as spinning on their axes. With NMR, we are performing experiments on the nuclei of atoms. If the atom spins are paired against each other, then the nucleus of the atom has no overall spin. In some atoms, the atom spins are unpaired and the nucleus possesses an overall spin. For example, hydrogen $^{1}H$ with an overall spin equal to 1/2 will have two possible orientations. Each orientation corresponds to the energy level. When a static magnetic field is applied, the energy levels split. The initial populations of the energy levels are determined by thermodynamics, as described by the Boltzmann distribution. It means that the lower energy level will contain slightly more nuclei than the higher level. It is possible to excite these nuclei into the higher level with electromagnetic radiation and observe exchanges between the energy levels.

NMR experiments can be carried out using continuous waves (CW) or pulsed NMR techniques. CW NMR remained the method of choice as a research tool for about 25 years, but pulsed NMR now completely dominates nuclear magnetic spectroscopy. In geophysics, CW NMR can be used only in a laboratory for investigating samples, and the MRS method is a pulsed NMR (Farrar and Becker 1971). The measuring procedure comprises irradiation of the investigated volume by a pulse of electromagnetic energy and when the electromagnetic radiation is cut off, to observe how the spin system returns to the previous state with the lower energy level containing more nuclei. In a static magnetic field, the frequency of radiation needed to excite the nuclei is determined by the difference in energy between the energy levels. It is equal to the Larmor frequency. In a fixed static magnetic field, the Larmor frequency is a signature of each type of nuclei possessing an overall spin. Figure 2.1 shows three principal steps necessary for measuring the NMR signal.

doi:10.1088/978-0-7503-3155-5ch2  

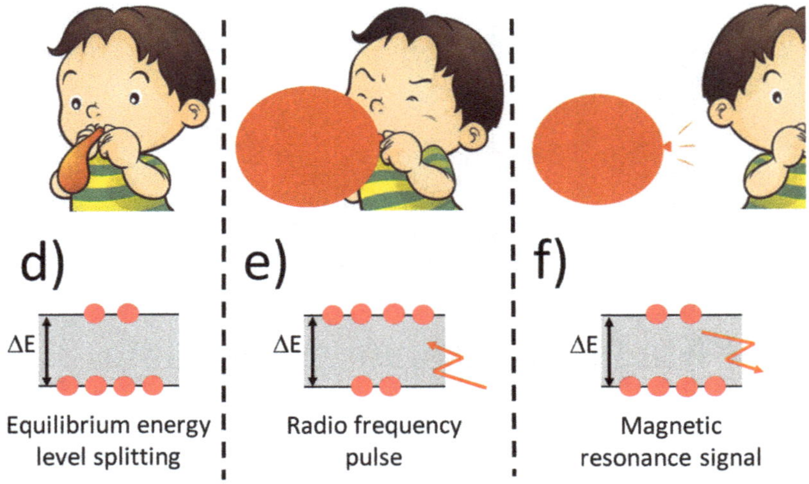


**Figure 2.1.** Schematic presentation of a pulsed NMR experiment.

First, let us represent the spin system by a balloon. Before we start our experiment, the balloon does not contain air and nothing happens (figure 2.1(a)). Then, we pump the balloon with air (figure 2.1(b)) and let the air escape (figure 2.1(c)). Escaping air produces a decaying whistle. The behavior of the spin system is similar. In equilibrium, the energy levels split so that the lower energy level contains more nuclei than the higher level (figure 2.1(d)). We irradiate the system with a pulse of a magnetic field oscillating at the Larmor frequency. The pulse changes populations of the energy levels (figure 2.1(e)). Then, the pulse is cut off and the spin system returns to the initial state with the lower energy level containing more nuclei (figure 2.1(e)). Returning to the equilibrium spin system generates a decaying magnetic field (the NMR signal) oscillating at the Larmor frequency. This magnetic field can be meaured and analyzed. It is important to note that without the spin system the NMR signal is equal to zero. The spin system that we use with the MRS method is groundwater and hence without groundwater the MRS signal is always equal to zero.

MRS is a large-scale pulsed NMR method operating in the Earth's magnetic field. The Earth's magnetic field is also called the geomagnetic field. The geomagnetic field is homogeneous, which allows for performing NMR experiments using a large measuring loop and investigating the subsurface in a volume as large as $150 \times 150 \times 150\ m^3$. MRS response is characterized by the following parameters: the initial amplitude $v_0$; the relaxation times $T_1$, $T_2$, $T_2^*$; and the phase shift $\Delta\phi$. The amplitude $v_0$ depends on the quantity of water below the MRS loop and the aquifers geometry. The relaxation times allow for estimating the saturated hydraulic conductivity of an aquifer formation. The phase shift is linked with the electrical resistivity of the subsurface. The Earth's magnetic field can be locally disturbed by rocks, which creates local heterogeneities and depending on these heterogeneities the magnetic resonance signal can be measured as the free induction decay (FID) signal or the

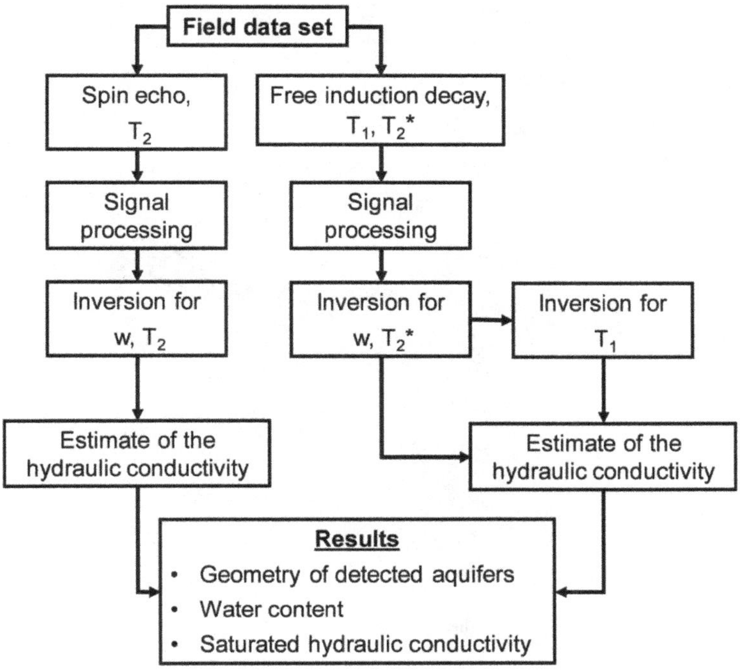


**Figure 2.2.** Flowchart of the interpretation procedure.

spin echo (SE) signal. Interpretation of MRS measurements comprises signal processing, forward modeling, inversion and an estimate of the hydraulic properties of the subsurface. Figure 2.2 presents a flowchart of the interpretation procedure.

MRS data processing, forward modeling and inversion depend on the measuring method. A spin echo data set means that measurements have been carried out in the area with inhomogeneous Earth's magnetic field. In this case, the relaxation time $T_2^*$ has little hydrological meaning. $T_2^*$ shows how much the Earth's magnetic field is disturbed and has no link to hydraulic properties of the media. Inversion of SE signals provides the geometry of water-bearing formations with the corresponding water content and an estimate of the saturated hydraulic conductivity. A free induction decay data set provides two relaxation times $T_1$ and $T_2^*$. In the undisturbed Earth's magnetic field, both of them can be used for estimating the saturated hydraulic conductivity. $T_2^*$ is linked with the inhomogeneity of the geomagnetic field and if the Earth's magnetic field is even slightly disturbed, then $T_2^*$ must be used with care. For this reason, $T_1$ is a more reliable parameter, but measuring $T_1$ is a more energy and time consuming procedure than measuring $T_2^*$.

MRS implementation is shown in figure 2.3. A large wire loop on the surface is used for generating an oscillating magnetic field. This loop, called the transmitting loop (Tx), defines the area investigated with one MRS station. The loop size may vary between $5 \times 5$ m$^2$ and $150 \times 150$ m$^2$, but usually between $50 \times 50$ m$^2$ and $100 \times 100$ m$^2$. The same loop can be used for measuring the MRS response from groundwater. In this case, the loop is called the coincident transmitting–receiving

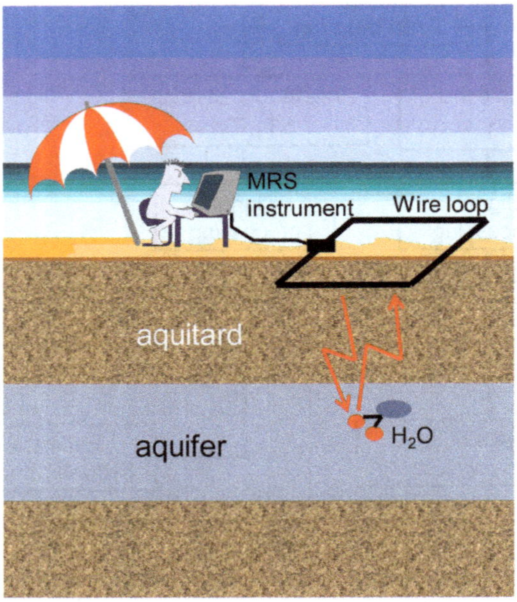


**Figure 2.3.** MRS field setup.

loop (Tx/Rx). Measurement of the MRS signal can also be done using one or more separated receiving loops Rx. The measuring procedure comprises the generation of one or more current pulse in the Tx loop and measuring the MRS response after each pulse in the Tx/Rx loop and/or one or more Rx loops.

All MRS instruments reported in the literature operate using similar algorithms and are built with similar principal units. A block diagram of typical MRS instruments is shown in figure 2.4.

The instruments may have one or more measuring loops. One loop is a transmitting loop (Tx) used for irradiating the spin system in the subsurface. This loop is made of a thick wire with the cross-section of 6 mm$^2$ or more. It is connected to a current generator. The current generator generates a pulse of current oscillating at the Larmor frequency. The maximum current used in MRS is up to 600 A with an output voltage up to 7.5 kV. The loop is tuned in resonance and current pulses are short, which reduces the power dissipated by the Tx loop and allows the use of standard car batteries for the power supply. Two batteries with a capacity of 75 Ah allow 4 to 8 hours of measuring. The power consumption depends on the field setup and measuring conditions. Usually, the same loop is used as a receiving loop (Rx). In this case, a Tx/Rx loop is connected through an electronic switch. The switch allows the fast commutation of the loop between the current generator and the amplifier. Amplified MRS signal is digitized by one or more analog-to-digital converters (ADC) and stored in the microprocessor memory. The microprocessor controls the entire instrument and the measuring procedure. Some part of data processing can be also done by the microprocessor. The microprocessor is connected to a common laptop PC used as the field computer. The field computer makes an interface

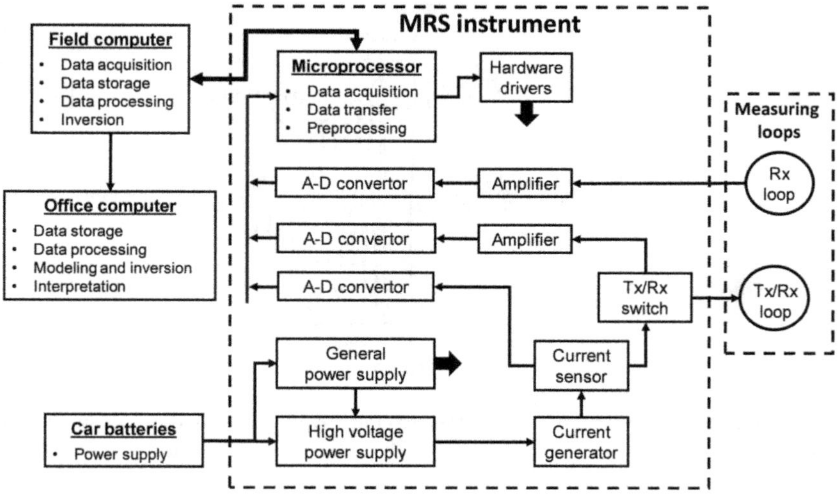


**Figure 2.4.** Block diagram of MRS instrument.

between the instrument and the operator. Interpretation of MRS data can be done in the field or in the office. We always recommend a rapid interpretation in the field for an express data analysis. In this case, errors in measured data can be corrected by performing additional measurements.

## 2.2 Measuring techniques

The simplest observation of the magnetic resonance phenomenon can be done using a one-pulse sequence shown in figure 2.5.

First, voltage induced in the Rx loop is recorded by the instrument. This voltage corresponds to an ambient electromagnetic (EM) noise. Then, a pulse of oscillating current is generated in the transmitting loop

$$i(t) = I_0 \times \cos(\omega_0 t) \tag{2.1}$$

where $I_0$ is the current amplitude and $f_0 = \omega_0/2\pi$ is the Larmor frequency. The pulse is characterized by the pulse moment

$$q = I_0 \times \tau_p \tag{2.2}$$

where $\tau_p$ is the pulse duration.

In the subsurface, the current creates an EM field. After the pulse is terminated, the instrument measures voltage induced in the Rx loop. If the voltage is equal to that measured before the pulse, then the magnetic resonance signal was not observed. Otherwise, we observe an exponentially decaying free induction decay signal (FID1) oscillating at the Larmor frequency. The MRS signal is characterized by the initial amplitude $v_0$, the phase shift $\Delta\phi$ and three relaxation times $T_1$, $T_2$ and $T_2^*$. The FID1 signal allows measuring $v_0$ and $T_2^*$

$$v_1(t) = v_0 \times \exp(-t/T_2^*) \times \sin(\omega_0 t + \Delta\varphi_1) \tag{2.3}$$

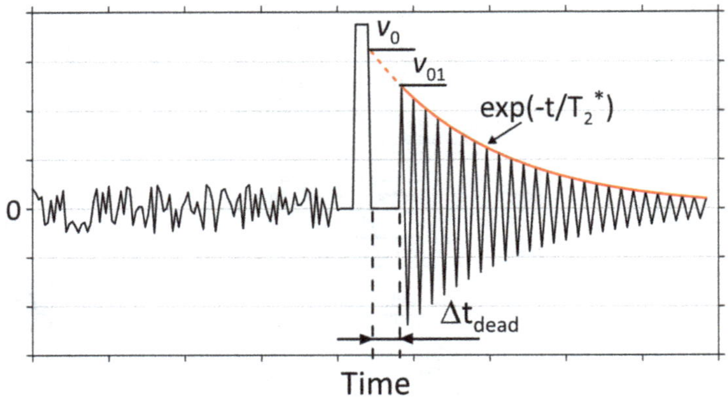


**Figure 2.5.** Schematic representation of a one-pulse measuring sequence. The red line shows the envelope of the MRS signal.

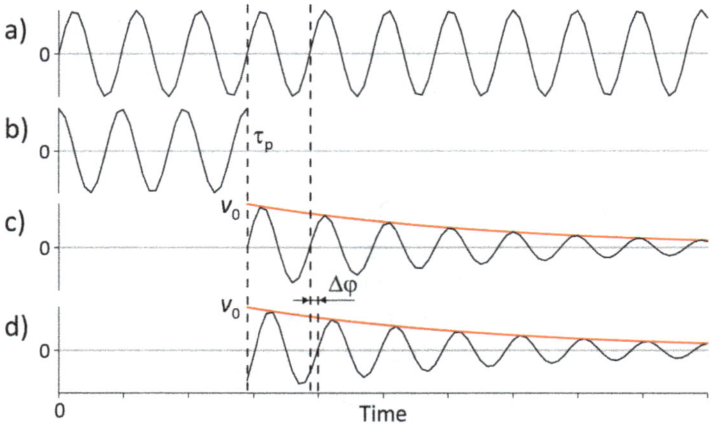


**Figure 2.6.** Measuring the phase shift: (a) the reference frequency; (b) the current pulse; (c) MRS signal without phase shift; (d) MRS signal with a positive phase shift.

where $v_0 \times \exp(-t/T_2^*)$ is the envelope. In practice, the FID1 signal is measured after an instrumental delay $\Delta t_{\mathrm{dead}}$ called the dead time

$$v_{01} = v_0 \times \exp(-\Delta t_{\mathrm{dead}}/T_2^*). \tag{2.4}$$

The signal phase is measured relative to the phase of the reference frequency (figure 2.6(a)). The current pulse is a cosines function of the reference frequency (figure 2.6(b)) and the MRS signal is a sinus function (figure 2.6(c)). Depending on measuring conditions, the phase of MRS signal may be shifted at the angle $\Delta\phi$. For example, figure 2.6(d) shows the case when $\Delta\phi > 0$.

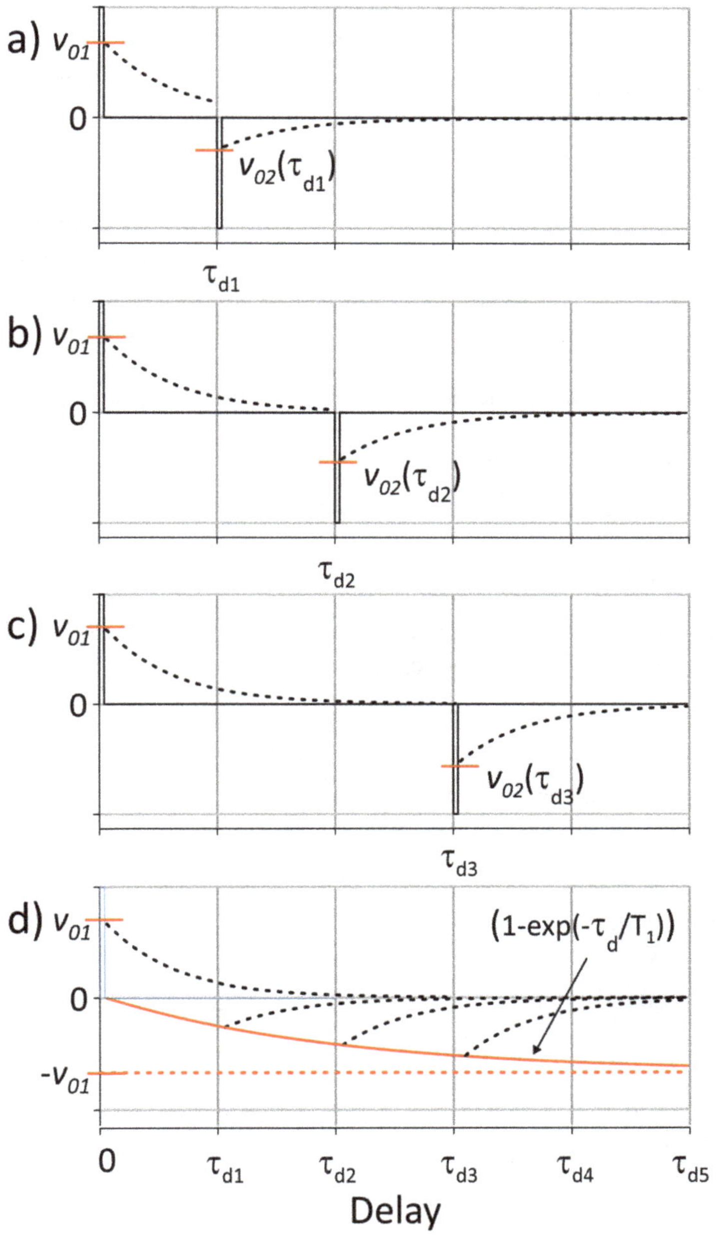


**Figure 2.7.** Measuring $T_1$ using a two-pulse sequence. Black dashed lines show free induction decay signals after the first pulse ($v_1$) and the second pulse ($v_2$). The solid red line shows the amplitude of the FID2 signal versus delay between the pulses.

For measuring $T_1$, one requires two pulses. In MRS, both pulses are set equal, but the first pulse is positive and the second pulse is negative (figure 2.7). The delay between the pulses $\tau_d$ is variable. The free induction decay signal (FID) is measured after the second pulse (FID2)

$$v_2(t) = v_{FID2} \times \exp(-t/T_2^*) \times \sin(\omega_0 t + \Delta\varphi_2) \tag{2.5}$$

where $v_{FID2}$ is the amplitude immediately after the second pulse. Similar to FID1, the amplitude of FID2 signals $v_{02}$ is measured after the dead time $\Delta t_{dead}$

$$v_{02} = v_{FID2} \times \exp(-\Delta t_{dead}/T_2^*). \tag{2.6}$$

Measurements of $T_1$ comprise measurements of $v_{02}$ for each value of the delay between the pulses $\tau_d$ (figures 2.7(a), (b), (c)). Figure 2.7(d) shows that the amplitude $v_{02}$ is negative and converges to the $-v_{01}$ with the rate given by $T_1$

$$v_{02}(\tau_d) = -v_{01} \times (1 - \exp(-\tau_d/T_1)). \tag{2.7}$$

For computing $T_1$, abs($v_{02}$) is plotted versus $\tau_d$ (figure 2.8) considering a shift $v_{shift}$ caused by unideal pulses.

$T_1$ is the solution of the following equation

$$v_{02}(\tau_d) = v_{shift} + \text{abs}(v_{01}) \times (1 - \exp(-\tau_d/T_1)). \tag{2.8}$$

In the homogeneous Earth's magnetic field, FID signals have an exponential envelope. However, the homogeneity of the Earth's magnetic field may be disturbed by rocks, which reduces the relaxation time $T_2^*$ and the MRS signal envelope becomes Gaussian instead of exponential one

$$v_1(t) = v_0 \times \exp\left(-\frac{1}{2}\left(\frac{t}{T_2^*}\right)^2\right) \times \sin(\omega_0 t + \Delta\varphi_1) \tag{2.9}$$

$$v_2(t) = v_{FID2} \times \exp\left(-\frac{1}{2}\left(\frac{t}{T_2^*}\right)^2\right) \times \sin(\omega_0 t + \Delta\varphi_2). \tag{2.10}$$

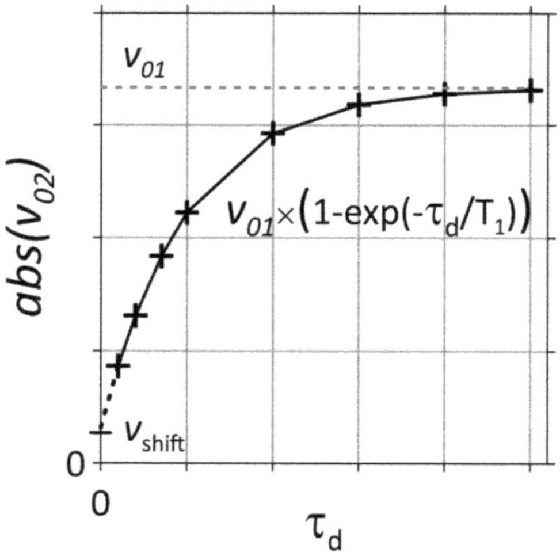


**Figure 2.8.** Amplitude of the FID2 signal versus delay between the pulses.

In the inhomogeneous Earth's magnetic field, the magnetic resonance signal can be measured using the spin echo method. For measuring spin echoes signals, one requires two pulses. In MRS, pulses have equal amplitudes, but the second pulse is twice as long. The amplitude of the second pulse can be positive or negative. The amplitude of the spin echo signal (se) has a Gaussian envelope and decays with the rate given by the relaxation time $T_2$ and the molecular diffusion

$$\mathrm{se}(t,\,\tau_d) = \mathrm{se}_0 \times \exp\left(-\frac{2\tau_d}{T_2} - \frac{2D\gamma^2G^2\tau_d^3}{3}\right) \times \exp\left(-\frac{1}{2}\left(\frac{t - 2\tau_d}{T_2^*}\right)^2\right) \tag{2.11}$$

where $\gamma$ is the proton's gyromagnetic ratio, $G$ is the spatial magnetic field gradient, $D$ is the diffusion coefficient.

Figure 2.9 shows spin echo signals as a function of the delay between the pulses. If $T_2^*$ is longer than the dead time of the instrument, then FID signals can also be observed after each pulse. The spin echo arrives after the second pulse at time $2\tau_d$ relative to the first pulse.

The amplitude of spin echo signals decreases for larger values of $\tau_d$. If the molecular diffusion can be neglected, then the decay is exponential

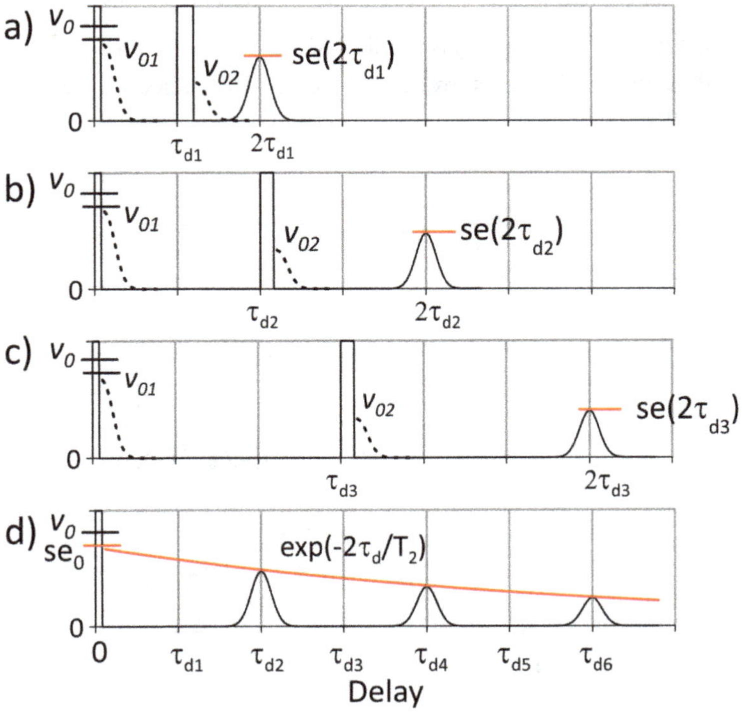


**Figure 2.9.** Measuring $T_2$ using a two-pulse spin echo sequence. Spin echo (solid line) and FID (dashed line) signals are shown considering different delays between the pulses (a), (b), (c). (d) Amplitude of the spin echo signal is shown neglecting the diffusion effect (red line).

$$\mathrm{se}(\tau_d) = \mathrm{se}_0 \times \exp\left(-\frac{2\tau_d}{T_2}\right). \tag{2.12}$$

Figure 2.9(d) shows the envelope of spin echoes (red line). A large volume investigated by MRS and unideal pulses cause the spin echo initial amplitude ($\mathrm{se}_0$) to be smaller in comparison with the FID1 amplitude $v_0$.

For computing $T_2$, the measured spin echo amplitude is plotted versus delay between the pulses. The delay $\tau_d$ cannot be set to zero and hence $\mathrm{se}_0$ is given by the extrapolation.

The effect of molecular diffusion in the inhomogeneous Earth's magnetic field enhances relaxation and accelerates the decay of the echo amplitudes for large delays $\tau_d$

$$\mathrm{se}(\tau_d) = \mathrm{se}_0 \times \exp\left(-\frac{2\tau_d}{T_2} - \frac{2D\gamma^2 G^2 \tau_d^3}{3}\right). \tag{2.13}$$

Visually, an echo envelope affected by the diffusion looks like exponential, but decaying faster than the exponential one (figure 2.11).

For example, figure 2.12 shows an exponential echoes envelope $\mathrm{se}(\tau_d)$ computed assuming the amplitude $\mathrm{se}_0 = 100$ nV and the decay time $T_2 = 300$ ms with the relaxation enhanced by diffusion (black crosses, $N$ values). These data were fitted by an exponential envelope $\tilde{\mathrm{se}}(\tau_d)$ ($\mathrm{se}_0 = 110$ nV, $T_2 = 221$ ms) neglecting diffusion (red line). The root-mean-square error (RMSE) is small ($\mathrm{RMSE} = \sqrt{\frac{1}{N}\sum_{i=1}^{N}(\mathrm{se}_i - \tilde{\mathrm{se}}_i)^2} = 2.8$ nV), but the initial echo amplitude is overestimated at 10% and the relaxation time is underestimated at 26%.

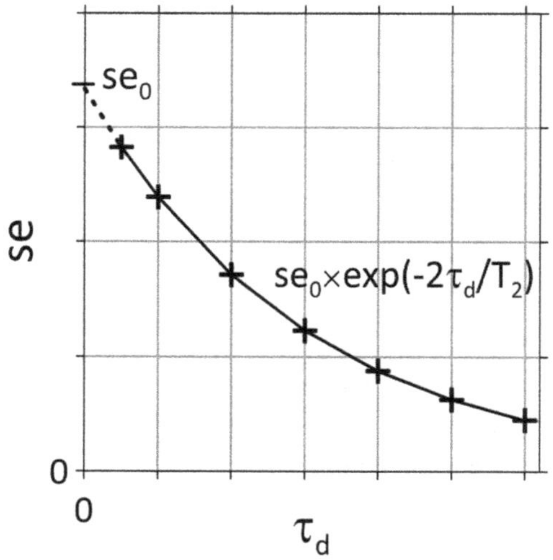


**Figure 2.10.** Amplitude of spin echo signal versus delay between two pulses is plotted neglecting the diffusion effect.

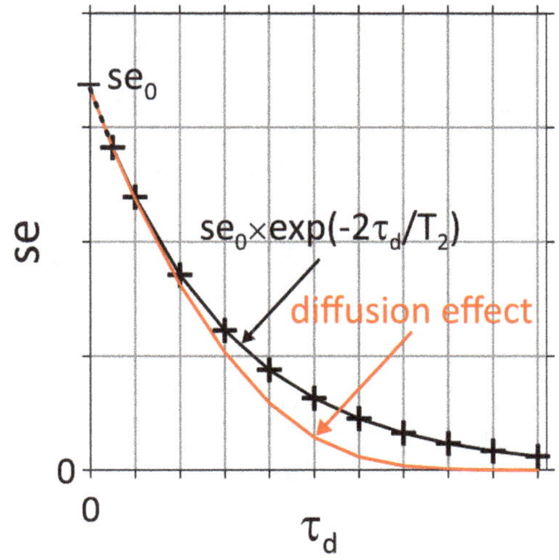


**Figure 2.11.** Effect of the enhanced relaxation on the echo amplitudes caused by the diffusion.

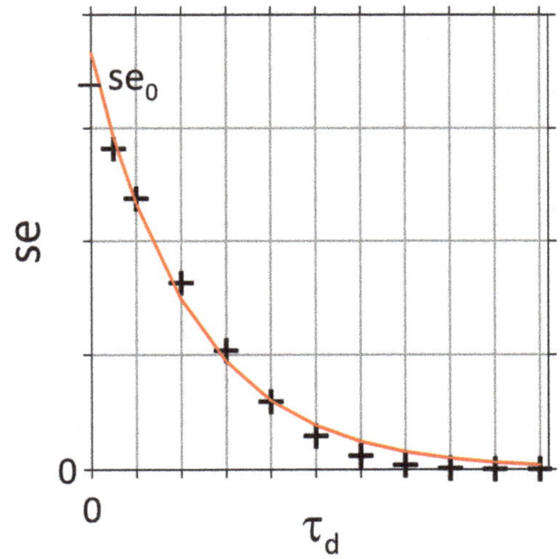


**Figure 2.12.** Spin echo amplitudes computed considering the diffusion effect (crosses) is plotted versus delay between the pulses. The red line shows an exponential function computed neglecting diffusion.

At the scale of MRS measurement, the diffusion coefficient $D$ and the field gradient $G$ are unknown, and it is convenient to fit the echoes envelope by introducing the diffusion factor $F_{\mathrm{dif}}$.

$$\mathrm{se}(\tau_d) = \mathrm{se}_0 \times \exp\left(-\frac{2\tau_d}{T_2} - F_{\mathrm{dif}}\tau_d^3\right) \quad (2.14)$$

where

$$F_{\mathrm{dif}} = (2D\gamma^2 G^2)/3. \quad (2.15)$$

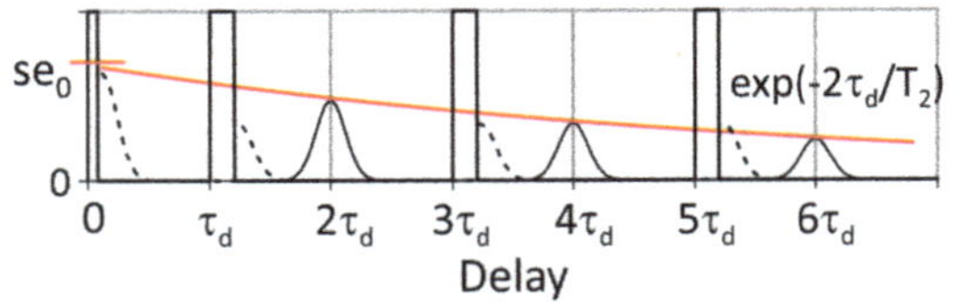


**Figure 2.13.** Measuring $T_2$ using the CPMG method. Pulses and spin echo (solid line) and FID (dashed line) signals are shown considering different delays between the pulses. The amplitude of the spin echo signals decays with the time constant $T_2$ (red line).

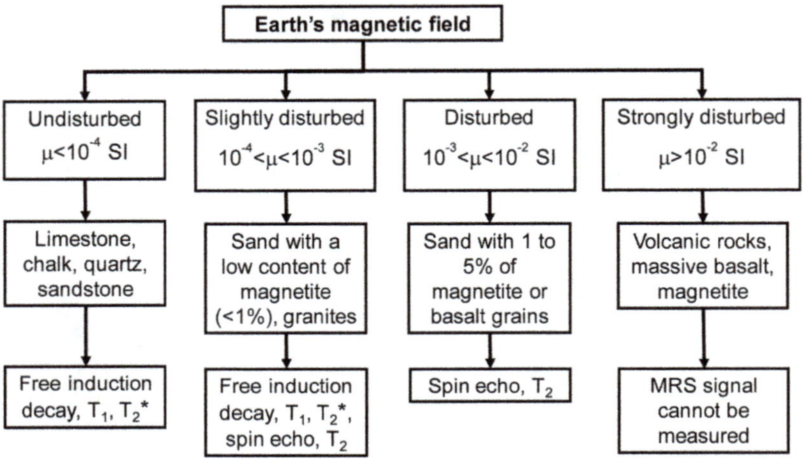


**Figure 2.14.** Selection of the measuring method as a function of the homogeneity of the Earth's magnetic field.

Equation (2.14) can be solved for $se_0$, $T_2$ and $F_{dif}$. While $se_0$ and $T_2$ are two parameters of the magnetic resonance signal we are looking for, $F_{dif}$ is applied to offset the diffusion effect.

In laboratory and borehole NMR instruments, the effect of diffusion is usually reduced by using multi-pulse measuring sequences. One of them, the Carr–Purcell–Meiboom–Gill sequence (CPMG) is shown in figure 2.13 (Farrar and Becker 1971). The CMPG method requires generating many pulses with the minimal delay $\tau_d$ limited by the duration of the FID and spin echo signals.

This method was adapted to MRS measurements (Grunewald and Walsh 2013), but the necessity of generating many powerful pulses in a short time interval meets instrumental limitations for mobile MRS instruments and CMPG measurements can be used only for investigating shallow targets in the inhomogeneous Earth's magnetic field.

Figure 2.14 suggests selection of the measuring method as a function of the homogeneity of the Earth's magnetic field that can be disturbed by surrounding rocks. Different rocks have different magnetic properties defined by the magnetic properties of the constituent mineral grains and crystals. The magnetic properties of the rock are characterized by the magnetic permeability $\mu$ and the remanent magnetization. The remanent magnetization acts as a permanent magnet and the magnetic permeability is the degree of magnetization that a material obtains in response to an applied magnetic field. For using MRS, it is often sufficient to know the magnetic permeability. Some rocks (chalk, limestone) do not contain magnetic minerals ($\mu < 10^{-5}$ SI) and some others (magnetite, basalt) contain a lot ($\mu > 10^{-2}$ SI). When soils and rocks have a low magnetic permeability ($\mu < 10^{-4}$ SI), only FID measurements are possible. In disturbed geomagnetic field ($10^{-3}\mu < 10^{-2}$ SI), only spin echo measurements are possible. In slightly disturbed geomagnetic field, both FID and SE signals can be observed. When the Earth's magnetic field is strongly disturbed by geological formations, the MRS method cannot be applied. In general, measurements of the FID signal allow a larger depth of investigation and the FID1 signal $v_{01}$ is larger than the spin echo signal $se(2\tau_{d\ min})$, which allows better signal-to-noise ratio. Thus, FID measurements should be preferred except in cases when the Earth's magnetic field is disturbed and FID signals cannot be observed.

## 2.3 Short MRS signals

We define short MRS signals as rapidly decaying signals with $T_2^* < 40$ ms. In the undisturbed Earth's magnetic field, short MRS signals are also characterized by short relaxation times $T_2 \approx T_2^*$ and $T_1 \approx 1.6 \times T_2^*$. But if the Earth's magnetic field is disturbed by rocks, then short MRS signals may have $T_1$, $T_2 >> T_2^*$.

For measuring, it is important to use current pulses with the bandwidth larger than that of the MRS signal. Both the pulse and the MRS signal have limited bandwidths centered at the Larmor frequency. Figure 2.15(a) schematically shows spectra of a spin system and of a pulse. For this demonstration, we assume the pulse spectrum to be an ideal rectangle. If the pulse bandwidth is larger than the signal bandwidth, then the MRS signal can be fully excited by the magnetic field produced

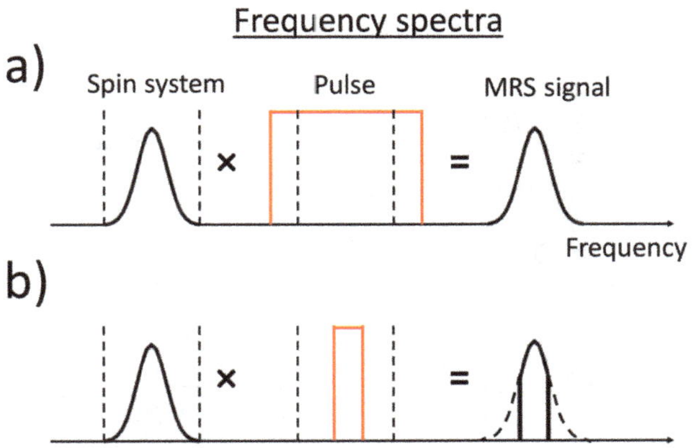


**Figure 2.15.** Radiofrequency pulse as a band-pass filter: (a) the bandwidth of the pulse (red line) is large relative to that of the spin system (black solid line); (b) the bandwidth of the pulse is narrow relative to that of the spin system.

by the pulse. However, if the pulse bandwidth is narrow relative to the signal bandwidth, then the response from the spin system will be degraded (figure 2.15(b)).

Figure 2.16 shows normalized frequency spectra of exponentially decaying signals centered at the Larmor frequency. For this demonstration, we assume three short MRS signals with different values of $T_2^*$.

Figure 2.17 shows spectra of the rectangular pulses of three different lengths. These graphs demonstrate the link between the bandwidth and the pulse length.

Figures 2.18, 2.19 and 2.20 show spectra of the rectangular pulses and exponential signals with the relaxation time $T_2^*$ of 10, 20 and 40 ms, respectively. As a rule of thumb, the presented graphs suggest using the pulse length equal or shorter than the relaxation time $T_2^*$ ($\tau_p \leqslant T_2^*$).

MRS is a pulsed NMR method and it supposes the use of short pulses. But the use of short pulses requires a large current in the transmitting loop. Otherwise, the depth

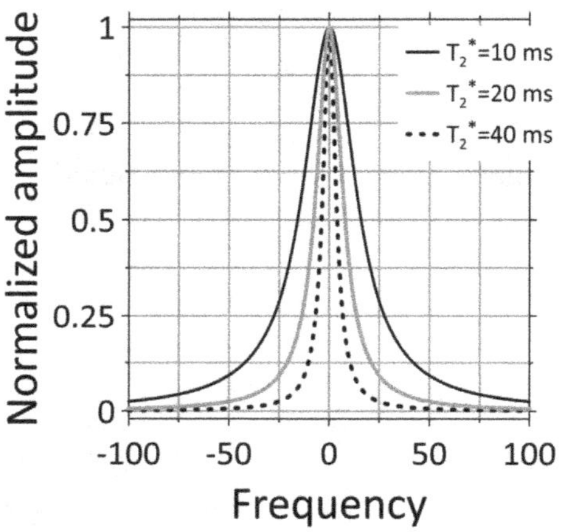


**Figure 2.16.** Normalized spectra of exponentially decaying signals centered at the Larmor frequency computed assuming different time constants.

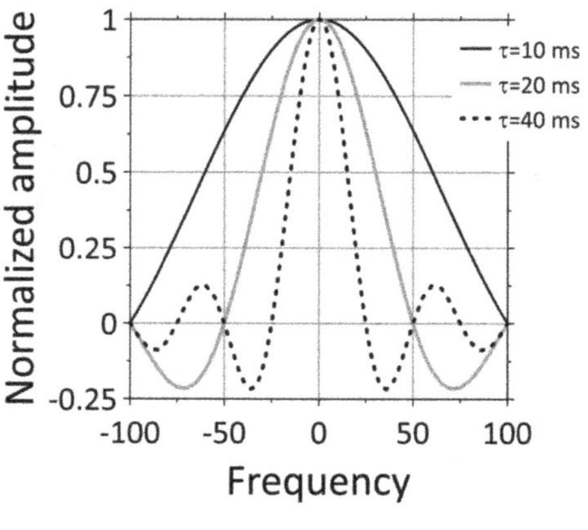


**Figure 2.17.** Normalized spectra of radiofrequency rectangular pulses of different lengths centered at the Larmor frequency.

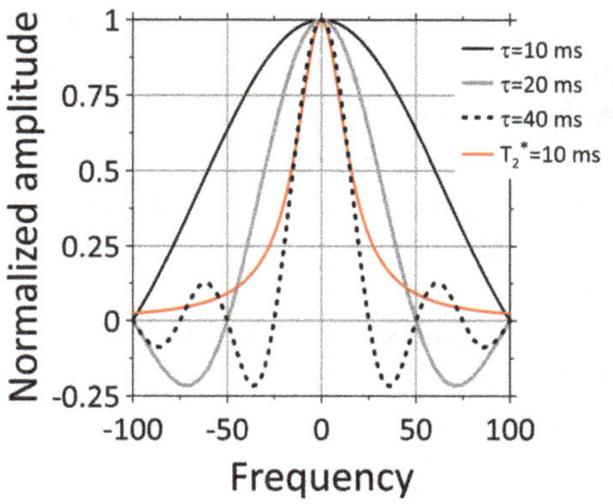


**Figure 2.18.** Normalized spectra of the radiofrequency pulses centered at the Larmor frequency and of an exponentially decaying signal with $T_2$* = 10 ms.

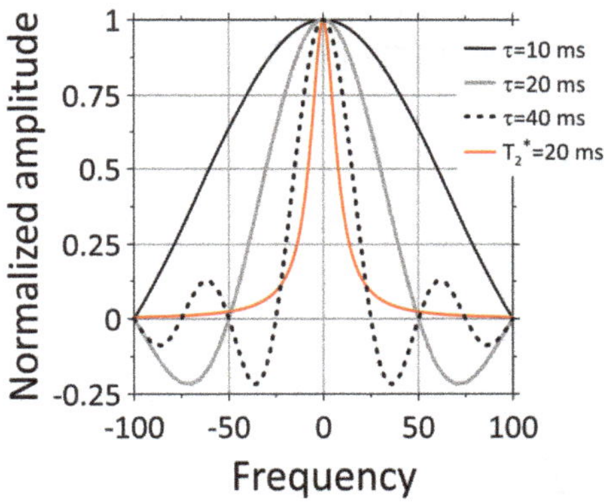


**Figure 2.19.** Normalized spectra of the radiofrequency pulses centered at the Larmor frequency and of an exponentially decaying signal with $T_2$* = 20 ms.

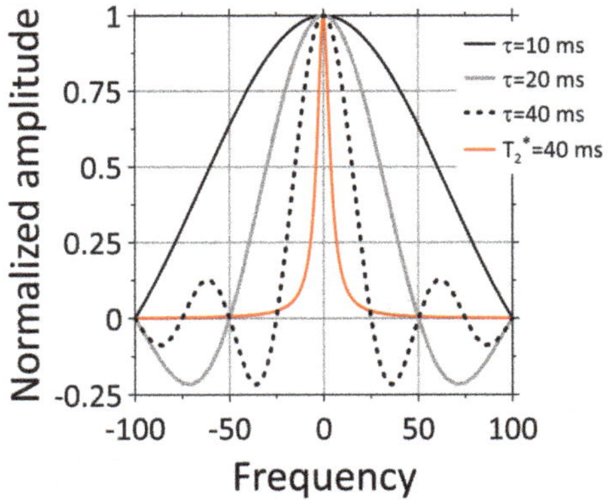


**Figure 2.20.** Normalized spectra of the radiofrequency pulses centered at the Larmor frequency and of an exponentially decaying signal with $T_2$* = 40 ms.

of investigation will be reduced. Figure 2.21 shows the maximum depth of investigation versus pulse moment computed considering a one-m-thick layer of bulk water and the threshold of the signal detection of 5 nV. For computing, we use an 80-m-side coincident Tx/Rx loop, a non-conductive subsurface (the half-space of 500 Ω·m) and the Earth's magnetic field of 47 000 nT (inclination of 65°N). The pulse moment $q$ that defined the depth of investigation is a product of the current in the loop and pulse duration ($q = I_0 \times \tau_p$).

In pulsed NMR, pulses are set much shorter than the relaxation times ($\tau_p << T_1$, $T_2$, $T_2^*$). This allows for neglecting relaxation processes during the pulse. If the pulse length is equal or longer than the relaxation times ($\tau_p \geqslant T_1$, $T_2$, $T_2^*$), then measurements can be seen as a CW NMR experiment. The NMR signal generated after the pulse cut off has an amplitude corresponding to a transient response of the spin system after the turn on of the radiofrequency field. Short pulses applied for MRS measurements reduce the depth of investigation and pulses are often set as long as possible ($\tau_p \approx T_2^*$). In this case, the relaxation during the pulse (RDP) can affect measured amplitude of the MRS signal. Figure 2.22 shows two exponentially decaying signals following the pulse ($\tau_p$) and the dead time ($\Delta t_{\text{dead}}$). The first signal is

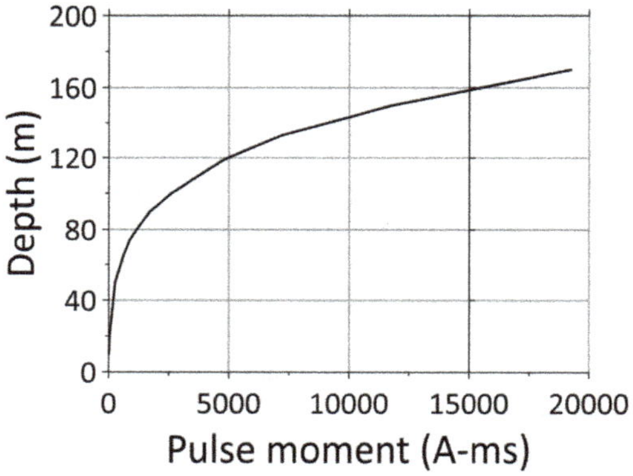


**Figure 2.21.** The maximum depth of water detection versus pulse moment computed considering an 80-m-side square loop coincident Tx/Rx loop.

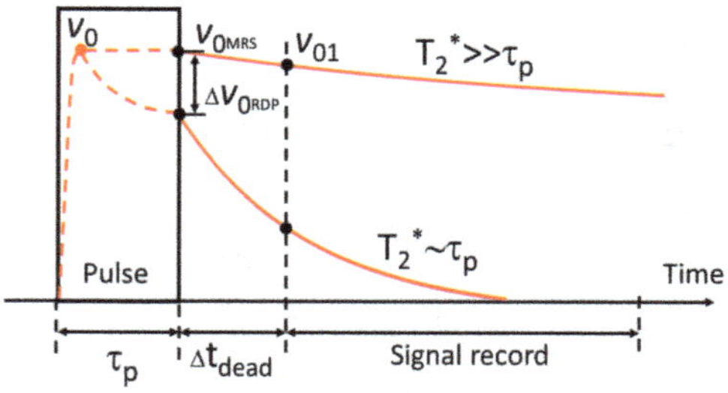


**Figure 2.22.** Relaxation of a short MRS signal during the pulse.

long relative to the pulse ($\tau_p << T_2^*$) and the relaxation during the pulse is negligible. The amplitude ($v_{01}$) and the relaxation time ($T_2^*$), both measured after the dead time, allow computing the initial amplitude ($v_0 \approx v_{0\,\mathrm{MRS}}$). The second signal is short ($\tau_p \approx T_2^*$) and the amplitude after the pulse measured with the MRS instrument is smaller than the initial amplitude ($v_0 = v_{0\,MRS} + \Delta v_{0\,RDP}$). $\Delta v_{0\mathrm{RDP}}$ depends on the pulse duration $\tau_p$, the relaxation times $T_1$ and $T_2^*$, and on the frequency offset ($\Delta\omega = \omega_0 - \omega$) between the current frequency ($\omega$) and the center frequency of the MRS signal ($\omega_0$). In MRS, the uncertainty in measuring the initial amplitude of the MRS signal ($v_0$) caused by $\Delta v_{0\mathrm{RDP}} \neq 0$ is neglected. The corresponding uncertainty in the water content estimate is a part of the total uncertainty in MRS results. This uncertainty can be reduced by experimental calibrations.

Let us assume short pulses ($\tau_p \leqslant T_2^*$) that allow exciting the entire spin population and neglect the relaxation during the pulse. Under these assumptions, the capacity of MRS measuring short FID signals is limited by the dead time of the instrument. During the dead time, the initial amplitude $v_0$ can be computed by extrapolation $v_0 \approx v_{0\mathrm{MRS}} = v_{01} \times \exp(\Delta t_{\mathrm{dead}}/T_2^*)$. If $\Delta t_{\mathrm{dead}} << T_2^*$, then the error in measuring $T_2^*$ has a little effect on the $v_0$ estimate. However, if $\Delta t_{\mathrm{dead}} \geqslant T_2^*$, then the error caused by an erroneous estimate of $T_2^*$ is amplified, which may have an effect on the accuracy of the amplitude estimate.

If the Earth's magnetic field is not homogeneous and $T_2^* << T_2$, then the spin echo procedure allows measuring signals shorter than the dead time because of the echo arrival is delayed relative to the pulse cut off. The first spin echo has the amplitude $\mathrm{se}_{01}$ and arrive at time $2\tau_d$. The initial amplitude of the spin echo $\mathrm{se}_0$ can be computed by extrapolation $\mathrm{se}_0 = \mathrm{se}_{01} \times \exp(2\tau_d/T_2)$. The minimum value of the echo separation ($\tau_{d\,\min}$) is selected so that both the FID and the SE signals fully decay during $\tau_{d\,\min}$. The MRS signal with the Gaussian envelope decreases by a factor of ten after $2.14 \times T_2^*$. For measuring, SE signal should be separated from the FID signal arriving after the second pulse (FID2), which yields $\tau_{d\,\min} \approx 2 \times T_2^*$. The extrapolation interval ($2\tau_{d\,\min} \approx 4 \times T_2^*$) imposes the measuring assumption $T_2 >> 4T_2^*$. If $T_2 \approx 4T_2^*$, then the error caused by an erroneous estimate of $T_2$ is amplified, which has an effect on the accuracy of the spin echo amplitude estimate. If $T_2 < 4T_2^*$, then the spin echo signal will be disturbed or even may not exist.

Modern MRS instruments allow measuring magnetic resonance signals with the minimum dead time $\Delta t_{\mathrm{dead}} = 5$ ms. The pulse length can be set between 5 and 50 ms and the current in the transmitting loop up to 500 A. Assuming the limit for the FID signal $T_{2\,\min}^* \approx \Delta t_{\mathrm{dead}}$, we get the pulse length $\tau_{p\,\min} = T_{2\,\min}^* = 5$ ms and consequently the eight fold reduction of the pulse moment $q$ relative to the pulse length of 40 ms, which is typically used with MRS. As shown in figure 2.21, smaller pulse moment will correspondingly reduce the maximum depth of investigation.

In practice, the reduced depth of investigation, a low vertical resolution, a poor signal-to-noise ratio and difficulties of accurate modeling of physical processes taking place on the grain surface limit the shortest signals measured with MRS at $T_{2\,\min}^* \approx 30$ ms.

## 2.4 Imaging equations

We assume an MRS field setup comprising a coincident transmitting/receiving loop on the ground surface and the FID1 signal measured using a one-pulse sequence. The loop is energized by a pulse of alternating current $i(t) = I_0 \cos(\omega t)$ with the frequency $\omega \approx \omega_0$. The frequency of the current is set close to the Larmor frequency of the hydrogen protons in the geomagnetic field

$$f_0 = \omega_0/2\pi = \gamma B_0/2\pi \tag{2.16}$$

where $B_0$ is the geomagnetic field and $\gamma$ the gyromagnetic ratio of the protons. The pulse is characterized by the pulse moment

$$q = I_0\tau_p \tag{2.17}$$

where $I_0$ and $\tau_p$ are the amplitude and duration of the pulse, respectively.

The pulse creates in the subsurface an oscillating magnetic field with the transverse to the Earth's magnetic field component $B_1(\mathbf{r})$ and the nuclear magnetization of protons rotates at the flip angle

$$\theta_p = 0.5\gamma I_0^{-1} B_1 q. \tag{2.18}$$

One sounding comprises measuring MRS signals applying pulses with different values of the pulse moment. The amplitude of MRS signal can be computed using the following equation

$$v_0(q) = I_0^{-1} M_0 \omega_0 \int_V e^{j\varphi_0(\mathbf{r})} B_1 \sin(\theta_p) w(\mathbf{r}) dV \tag{2.19}$$

where $M_0$ is the nuclear magnetization of the protons, $0 \leqslant w(\mathbf{r}) \leqslant 1$ the water content, and $\mathbf{r} = r(x, y, z)$ the coordinate vector. Under near resonance conditions, MRS signal generated by a volume $dV$ has a phase shift caused by the electrical resistivity of the rock

$$\varphi_0(\mathbf{r}) = 2 \tan^{-1}(\mathrm{Im}(B_1)/ \mathrm{Re}(B_1)). \tag{2.20}$$

The MRS signal oscillates at the Larmor frequency and exponentially decays with the time constant $T_2^*$

$$v_1(q, t) = v_0(q) \times e^{-t/T_2^*(q)} e^{j(\omega_0 t + \varphi_0)}. \tag{2.21}$$

For measuring $T_1$, we use two pulses with the variable delay $\tau_d$ inbetween. The relaxation time and the phase of the FID2 signal are not used for interpretation of MRS measurements. After the second pulse, the amplitude of the MRS signal $v_2$ is given by the following equation

$$v_{02}(q, \tau_d) = -v_{01}(q) \times (1 - e^{-\tau_d/T_1(q)}). \tag{2.22}$$

In the inhomogeneous geomagnetic field, the free induction signals decay very fast and it can be difficult or impossible to measure. In this case, spin echo measurements can be carried out. We consider two pulses. The pulse moment of the first pulse $q_1$

and of the second pulse $q_2$ so that $q_1 = q_2/2 = q$ and hence the flip angle is $\theta_{p1} = \theta_{p2}/2 = \theta_p$. Under near resonance conditions and neglecting relaxation, the spin echo amplitude can be calculated as

$$\mathrm{se}_0(q) = -I_0^{-1}M_0\omega_0 \int_V e^{j\varphi_0(\mathbf{r})}B_1 \sin^3(\theta_p)w(\mathbf{r})dV. \tag{2.23}$$

For measuring $T_2$, we use two pulses with a variable delay $\tau_d$ inbetween. The spin echo arrives after the second pulse at time $2\tau_d$ and its amplitude is

$$\mathrm{se}(q, \tau_d) = \mathrm{se}_0(q) \times \exp\left(-\frac{2\tau_d}{T_2(q)} - F_{\mathrm{dif}}\tau_d^3\right). \tag{2.24}$$

## 2.5 MRS and groundwater

Imaging equations used for interpretation of MRS measurements are presented in the previous section. These equations have been developed under the assumption that the total MRS signal is recovered. However, relaxation may reduce measured signal amplitudes and for interpretation of MRS measurements in terms of groundwater the loss of the signal has to be taken into account. The relaxation mechanisms are described by three relaxation time constants $T_1$, $T_2$, and $T_2^*$.

The spin–lattice relaxation time $T_1$ depends on the strength of the coupling between the spin system and the lattice. A number of different physical interactions can cause an energy exchange between the two systems. In general, a stronger coupling causes faster relaxation.

The spin–spin relaxation time $T_2$ characterizes the energy exchange between the spins inside the spin system. In general, $T_2 < T_1$. In porous media, the transverse relaxation time $T_2$ is also controlled by surface interactions. The $T_1/T_2$ ratio for a variety of porous sandstone and carbonate rocks (105 samples) was reported between 1 and 2.6 with the mean about 1.65 (Kleinberg *et al* 1993). The observed relaxation time $T_2^*$ includes the spin–spin relaxation and the Earth's magnetic field inhomogeneity $\Delta B_0$ caused by geological materials

$$1/T_2^* = 1/T_2 + \gamma\Delta B_0/2 \tag{2.25}$$

where $\gamma$ is the gyromagnetic ratio for protons. In the homogeneous Earth's magnetic field $\Delta B_0 = 0$ and $T_2^* = T_2$.

Figure 2.23 shows a pore space schematically divided into three principal zones: a surface layer of thickness $\lambda$, such that any spin that enters this layer relaxes with the time constant $T_{1S}$ (red band); a central part of the pore space with the bulk relaxation time $T_{1B}$ (green area); and the transitional zone inbetween (yellow band).

During measuring time, spins in the transitional zone travel between the surface and the central parts of the pore space changing many times the relaxation time for each spin from $T_{1B}$ to $T_{1S}$ and from $T_{1S}$ to $T_{1B}$. A great number of spins and their fast movement cause the mean relaxation time from the transitional zone varying so that $T_{1S} < T_{1\mathrm{trans}} < T_{1B}$. Considering the self-diffusion coefficient for water, the travel distance

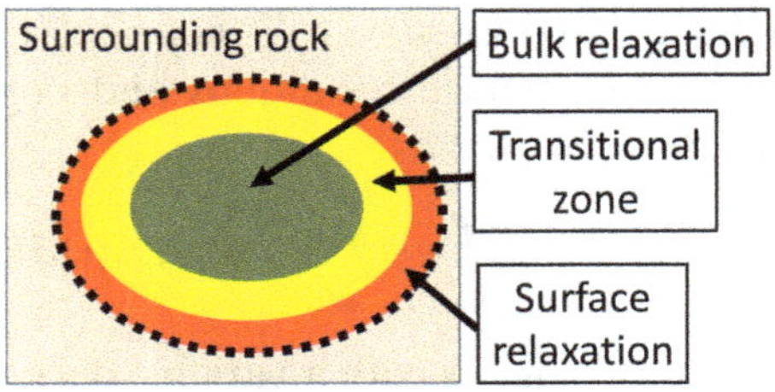


**Figure 2.23.** The pore space divided into three zones considering dominating relaxation mechanism: the surface relaxation (red), the bulk relaxation (green) and the transitions zone inbetween (yellow).

of spins can be estimated as about 100 μm $s^{-1}$. In pores larger than the travel distance, influence of the surface relaxation decreases. The total relaxation rate measured with MRS is a volumetric rate average of spins in the pore space. It can be written as

$$\frac{1}{T_1} = \frac{1}{T_{1B}} \times \left(1 - \frac{\lambda S}{V}\right) + \frac{\lambda S}{VT_{1S}}. \tag{2.26}$$

The bulk relaxation of water $T_{1B}$ ranges between 1 and 2.6 seconds. The surface relaxation $T_{1S}$ is much shorter and that of sandstone, for example, is of the order 0.1 to 1 ms. Considering that $T_{1B} >> T_{1S}$ we can write

$$\frac{1}{T_1} = \frac{1}{T_{1B}} + \frac{\lambda S}{VT_{1S}} = \frac{1}{T_{1B}} + \frac{\rho_1 S}{V} \tag{2.27}$$

where $\rho_1 = \frac{\lambda}{T_{1S}}$ is the surface relaxivity for $T_1$.

Similar arguments are also hold true for $T_2$

$$\frac{1}{T_2} = \frac{1}{T_{2B}} + \frac{\rho_2 S}{V} \tag{2.28}$$

where $\rho_2 = \frac{\lambda}{T_{2S}}$ is the surface relaxivity for $T_2$.

Equations (2.27) and (2.28) show that the relaxation times $T_1$ and $T_2$ are proportional to the pore size characterized by the volume to surface ratio $V/S$ and the surface relaxivity ($\rho_1 \neq \rho_2$). The surface relaxivity of porous materials is very variable. For example, reported values of sandstones show more than 100 fold difference ranging from 0.37 to 46 and $\rho_1$ of a quartz sand was reported about 0.02.

The $V/S$ ratio depends on the pore shape. For example, consider a sphere, a cylinder and a cone (figure 2.24). For simplicity, we assume that these objects are characterized by the radius of the footprint $r$. Figure 2.25 shows the corresponding $V/S$ ratio versus $r$. The value of $V/S$ ratio differs more than double the sphere and the cone. For all three objects $V/S$ increases proportionally to the object size.

The pore space of different geological materials may have different shapes. For example, let us consider a thin square box (figure 2.26) having the volume $V = a \times b \times h$ and the surface area $S = 2 \times (a \times b + a \times h + b \times h)$. Assuming $a >> h$, $b >> h$ and increasing the box size by increasing $a$ and $b$, we can get a large flat object with a small $V/S$ ratio.

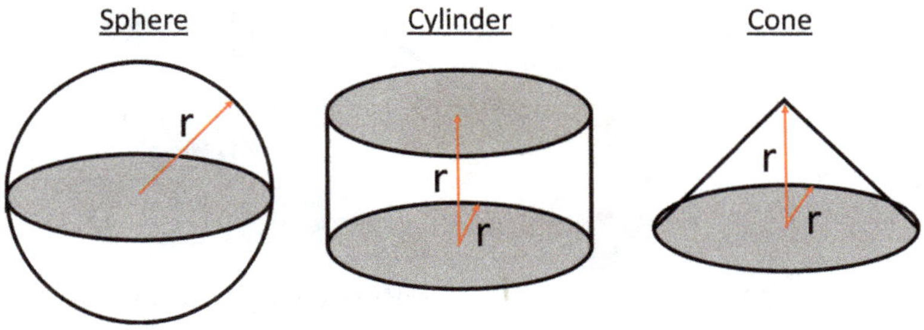


**Figure 2.24.** Common geometric objects used for calculating the volume to surface ratio characterized by the radius of the footprint.

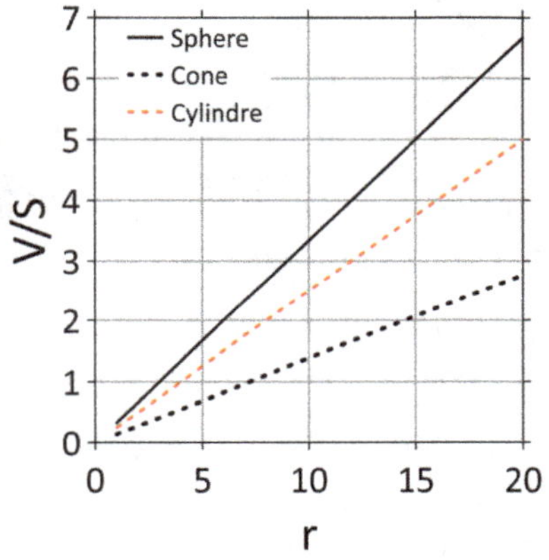


**Figure 2.25.** The volume to surface ratio (*V/S*) computed as a function of the object size characterized by the radius *r*.

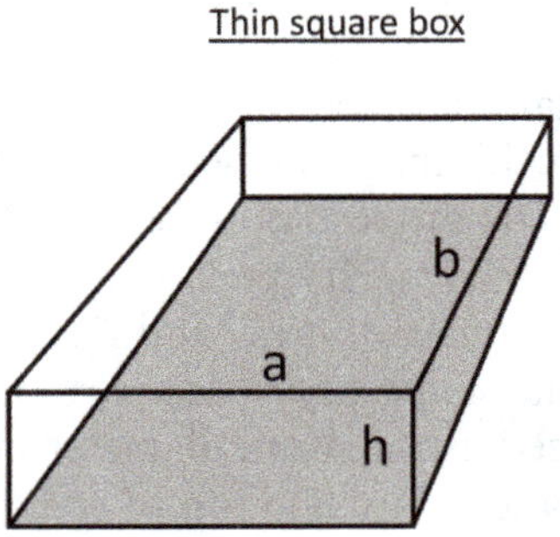


**Figure 2.26.** A thin square box with the dimensions *a»h* and *b»h*.

Another example shows that pores may be large, but the pore surface may not be smooth. The grain walls may have a very irregular shape (figure 2.27) and cause a small $V/S$ ratio.

Presented examples show that $V/S$ is a material dependent parameter and hence, the MRS estimation of the pore size is site-dependent.

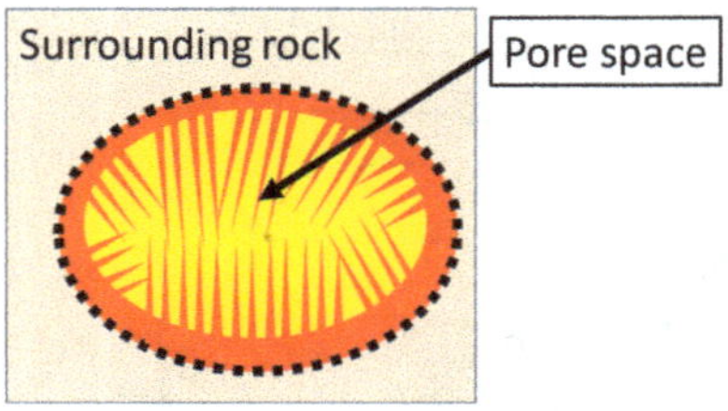


**Figure 2.27.** Example of a large pore with very developed surface area.

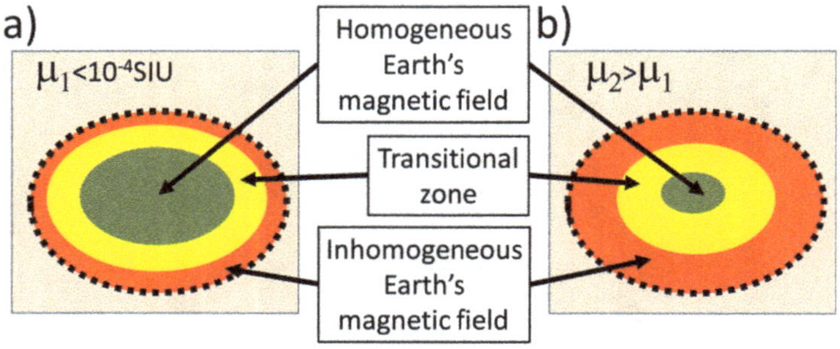


**Figure 2.28.** Homogeneity of the Earth's magnetic field disturbed by surrounding geological materials. (a) The magnetic permeability $\mu_1$. (b) The magnetic permeability $\mu_2$.

In different geological materials, pores may have different shapes, but surrounding rocks may also have different magnetic properties. Figure 2.28(a) shows the pore space surrounded by rocks with the magnetic permeability $\mu_1$, and figure 2.28(b) shows the rock with the magnetic permeability $\mu_2$ ($\mu_2 > \mu_1$). The Earth's magnetic field is disturbed by rocks and the area affected by the rocks depends on the $\mu$ value. This area can be much larger than the pore size. In some magnetic rocks like basalt, the Earth's magnetic field can be so much disturbed that MRS measurements are not possible.

Equation (2.25) shows that additionally to $T_2$, the relaxation time $T_2^*$ depends on the inhomogeneity of the Earth's magnetic field caused by surrounding rocks ($\Delta B_0$). $\Delta B_0$ is linked with the magnetic properties of the rock and is disconnected from the hydraulic properties of the media. It renders $T_2^*$ a less reliable parameter for estimating the pore size in comparison to $T_1$ and $T_2$.

In practice, measurements of the pore space are more complex than measurements of the grain size defined as the mean diameter of grains. In 1922, Chester K Wentworth published the canonical definition of sediment grain sizes (Wentworth 1922) presented in table 2.1.

Wentworth developed his classification assuming that in a homogeneous geological formation, every particular grain and every particular pore may have their individual size and shape. But parameters averaged over a grand volume can characterize the entire formation. MRS is a large-scale method and provides results averaged over the large area. Thus, the Wentworth grain size classification is consistent with the MRS method. MRS was developed in early 1980 and initially only $T_2^*$ measurements have been used. In non-magnetic materials, $T_2^*$ is close to

**Table 2.1.** Sediment grain size classification defined by Wentworth (1922).

| Wentworth size class | Grain size (μm) |
|---|---|
| Very coarse sand | 1000–2000 |
| Coarse sand | 500–1000 |
| Medium sand | 250–500 |
| Fine sand | 125–250 |
| Very fine sand | 63–125 |
| Coarse silt | 31–63 |
| Medium silt | 15.6–31 |
| Fine silt | 7.8–15.6 |
| Very fine silt | 3.9–7.8 |
| Clay | 0.02–3.9 |

**Table 2.2.** Empirical relationship between the relaxation time $T_2^*$ measured with MRS and the grain size (aggregate class) of unconsolidated sediments defined by Schirov *et al* (1991).

| Schirov aggregate class | $T_2^*$ (ms) |
|---|---|
| Sandy clay | < 30 |
| Clayey sand, very fine sand | 30–60 |
| Fine sand | 60–120 |
| Medium sand | 120–180 |
| Coarse and gravely sand | 180–300 |
| Gravel deposits | 300–600 |
| Surface water bodies | 600–1500 |

$T_2$ and $T_2^*$ can be used for characterizing the $\frac{V}{\rho_2 S}$ ratio linked with the grain size. In 1991, Michael Schirov reported an empirical correlation between the aggregate class and the relaxation time $T_2^*$ measured with MRS (Schirov *et al* 1991). The data set corresponds to numerous field studies of non-magnetic unconsolidated sediments carried out in different parts of the former USSR (table 2.2).

Figure 2.29 summarizes the classifications of Wentworth and Schirov and shows the empirical relationship between the relaxation time $T_2^*$ and the grain size of the water-saturated unconsolidated sediments.

MRS performance is site-dependent and even a small percentage of magnetic particles in the subsurface can have a large effect on the capacity of MRS to detect groundwater. The maximum sensibility corresponds to unconsolidated non-magnetic sediments characterized by the relaxation times $T_1$ and $T_2^* \approx T_2$. Measurements and interpretation of short MRS signals is challenging and the majority of MRS applications are focused on aquifer investigations considering the groundwater detection threshold of $T_2^* \approx 30$ ms. Based on the Wentworth and Schirov

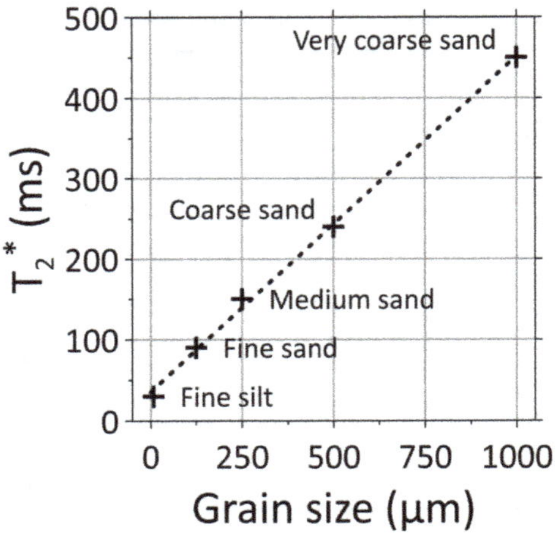


**Figure 2.29.** The relationship between the grain size and the relaxations time $T_2$* summarizing classifications developed by Wentworth (1922) and by Schirov *et al* (1991).

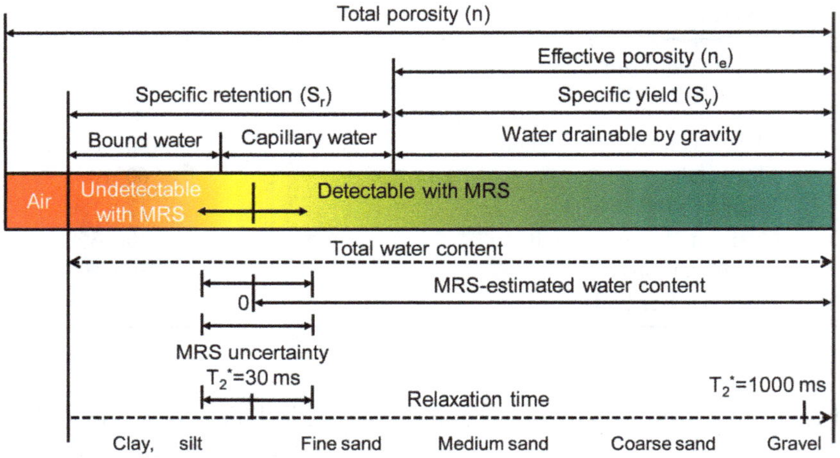


**Figure 2.30.** Hydrogeological classification of groundwater in the relationship with the relaxation time $T_2$* showing the MRS capacity to detect groundwater.

classifications, groundwater can be linked with the MRS capacity of water detection as shown in figure 2.30.

In soils and rocks, the total porosity shows the amount of empty space existing between the grains of minerals. The empty space can be occupied by air (/gases) and water (/liquids). The total water content is the fraction of the total rock filled with liquid water. This is a fraction between 0 and 1, but it must be less than or equal to the total porosity. Bound water is a few molecules thick layer of water surrounding mineral surfaces. A strong attraction between the water molecules causes much lower mobility than the rest of the water in the soil. The effective porosity can be thought of as the volume of pore space that will drain under the influence of gravity.

The effective porosity is close to the specific yield, which is defined as the volume of water released under gravitational force from an unconfined aquifer per unit surface area of aquifer per unit decline of the water table. The effective porosity and the specific yield can be thought of as equivalent for most non-technical purposes. The specific retention corresponds to the volume of water that a rock or soil will hold against the pull of gravity. A layer of water between the bound water and drainable water corresponds to capillary water. In clay and silt, the surface relaxivity is the principal relaxation mechanism and MRS signals measured in these formations are short. The threshold of the signal detection is located between a silt and fine sand. However, the capacity of MRS instruments measuring MRS signals and hence the capacity of the water detection depends not only on the grain size but also on the surface relaxivity and the magnetic permeability of soil particles affecting the relaxation times. In different geological materials, the threshold $T_2^* = 30$ ms may correspond to different grain sizes and MRS will show different water content in hydrologically identical materials. This is the reason of the uncertainty of measuring the water content with MRS. In coarse materials, contribution of the surface relaxivity is small and MRS is able to measure the water content with better accuracy.

The Wentworth grain size classes can be combined with the saturated hydraulic conductivity ($K_s$) predicted by the HYDRUS-1D software package. Figure 2.31 shows an approximate relationship between $K_s$ and the grain size.

The grain size can be replaced by the corresponding $T_2^*$ values reported by Schirov (figure 2.32). This empirical relationship allows predicting $K_s$ using $T_2^*$ measurements.

$$K_s = 1.89 \times 10^{-6} \times (T_2^*)^4. \quad (2.29)$$

Under favorable condition, MRS allows for estimating the grain size, but the hydraulic conductivity is also dependent on the pore connections termed in petrophysics as pore throats. Figure 2.33(a) shows two large pore volumes connected by thin channels. MRS characterizes the pore volume but water flows through thin connections, which renders the MRS predicted hydraulic conductivity erroneous.

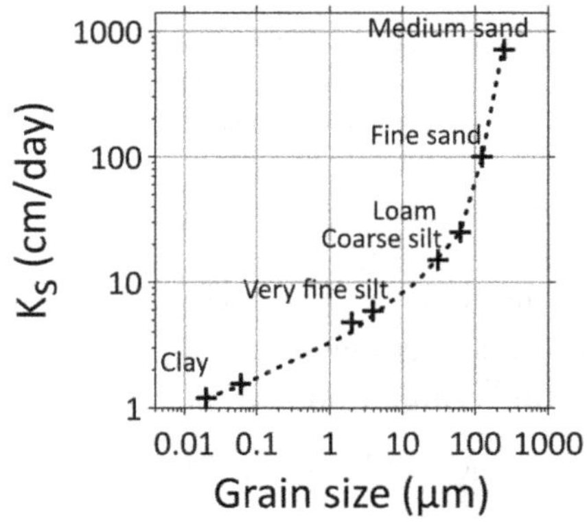


**Figure 2.31.** Approximate relationship between the grain size of unconsolidated sediments (Wentworth classification) and the saturated hydraulic conductivity predicted by the HYDRUS-1D software.

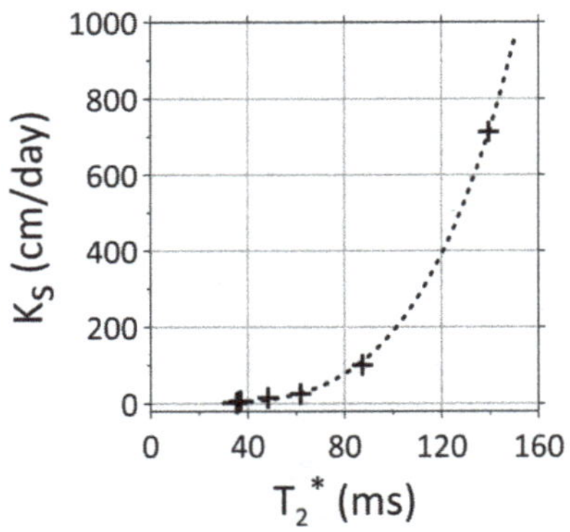


**Figure 2.32.** The relationship between the relaxations time $T_2$* corresponding to the experimental classifications of unconsolidated sediments developed by Wentworth and Schirov and the saturated hydraulic conductivity predicted by HYDRUS-1D software. The dashed line shows a power fit of the experimental data.

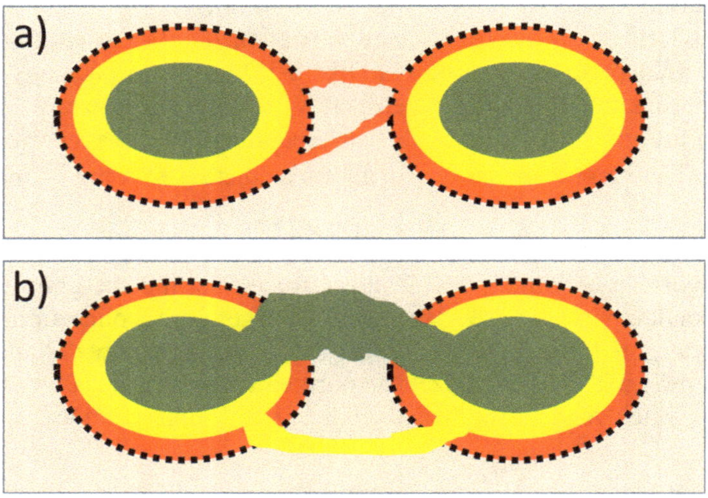


**Figure 2.33.** The capacity of MRS to predict the hydraulic conductivity in function of hydraulic connections between the pores: (a) connection channels (pore throats) are much smaller than the pore volume; (b) size of the connection channels is comparable with the pore volume.

Figure 2.33(b) shows two pores with the size of connections as large as the pore size. Here, MRS prediction will be correct.

The relationship shown in figure 2.32 can be used for prediction of the saturated hydraulic conductivity considering the sediments investigated by Schirov. But in petrophysics a more general formulation is used

$$K_s = C \times w^a \times T_1^b \tag{2.30}$$

where $w$ and $T_1$ the water content and the relaxation time measured with NMR and $C$, $a$ and $b$ are the empirical coefficients. If the spin echo measurements are used,

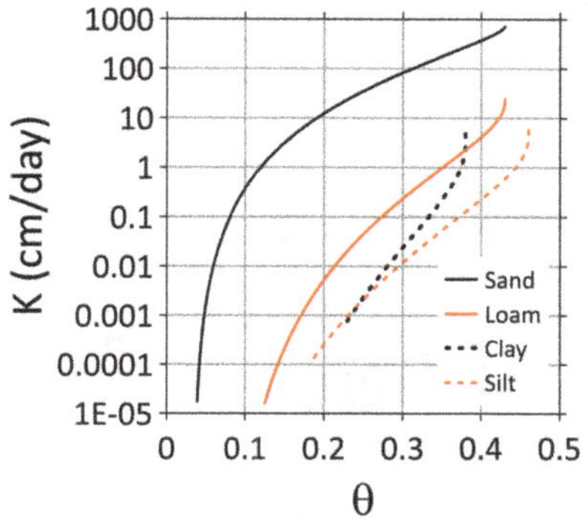


**Figure 2.34.** Examples of the unsaturated hydraulic conductivity versus water content corresponding to different unconsolidated sediments.

then $T_1$ is replaced by $T_2$. The empirical constant $C_p = 7.0 \times 10^{-9}$ was developed based on MRS studies of sand and clay aquifers and a simplified formula is often used for predicting the saturated hydraulic conductivity

$$K_s = C_p \times w \times T_1^2. \tag{2.31}$$

Initially, MRS was developed for investigating the water-saturated porous media with aquifers being the principal target. Later, it was found that MRS can be also applied for studying an unsaturated water flow in non-magnetic geological materials like limestone and chalk. Similar to the saturated media, MRS allows for estimating the water content and measuring the relaxation times $T_1$ and $T_2^*$. In the unsaturated media, the MRS relaxation times characterize thickness of the capillary layer but to link the relaxation times with the unsaturated hydraulic conductivity ($K$) is a more complex task than to link the relaxation times with the saturated hydraulic conductivity ($K_s$). In the unsaturated media, the hydraulic conductivity is a non-linear function of the water content as shown in figure 2.34. An estimate of the unsaturated hydraulic conductivity based on MRS measurements without knowledge of the $K(\theta)$ relationship corresponding to the investigated formation is hardly possible.

## Bibliography

Farrar E and Becker T 1971 *Pulse and Fourier Transform NMR, Introduction to Theory and Methods* (New York: Academic)

Grunewald E and Walsh D 2013 Multiecho scheme advances surface NMR for aquifer characterization *Geophys. Res. Lett.* **40** 6346–50

Kleinberg R, Farooqui S and Horsfield M 1993 T1/T2 ratio and frequency dependence of nmr relaxation in porous sedimentary rocks *J. Colloid Interface Sci.* **158** 195–98

Schirov M, Legchenko A and Creer G 1991 A new direct non-invasive groundwater detection technology for Australia *Explor. Geophys.* **22** 333–38

Wentworth C K 1922 A scale of grade and class terms for clastic sediments *J. Geol.* **30** 377–92

# Chapter 3

# Processing field records

## 3.1 Noise-contaminated MRS signal

Magnetic resonance sounding (MRS) signal is small and the signal-to-noise ratio (SNR) may often be poor. It justifies the presentation of different noise sources that can contaminate MRS records. Electromagnetic (EM) noise can be internal and external (ambient). The internal noise is generated by the measuring device and the external noise is induced in the measuring loop by external EM fields. Modern MRS instruments produce much smaller electronic noise than the ambient EM noise and we omit discussions about the amplifier design. The ambient EM noise can be generated by different sources in a wide frequency range, but we will consider only the frequency range approximately coinciding with that of the MRS signals (1–3 kHz).

Figure 3.1 shows the principal sources of the ambient (EM) noise. The ambient EM noise may be natural and man-made. Natural electromagnetic noise sources have existed since the origin of the Earth. A primary source of noise is given by the interaction of particles and waves coming from outer space with the magnetosphere plasma at various altitudes. Inside the ionosphere, atmospheric lightning discharges produce several propagating phenomena. The Earth's magnetic field is not only static but also includes slow time variations: secular, annual, diurnal; and faster variations caused by magnetic storms. Man-made noise is generated by human technologies implanted for power transmission and communications: communication and broadcasting systems, electric energy transport systems, automotive ignition, industry and instruments for general purposes and scientific appliances. In industrial areas or close to power lines, the man-made noise largely dominates the ambient noise. In rural areas, the natural noise is often the principal source of the interference. The natural noise is random and broad-band, and the man-made noise is concentrated in particular frequencies in association with technological applications. Both natural and man-made noises may contain short spikes overcoming the mean noise by a few orders of magnitude.

doi:10.1088/978-0-7503-3155-5ch3  

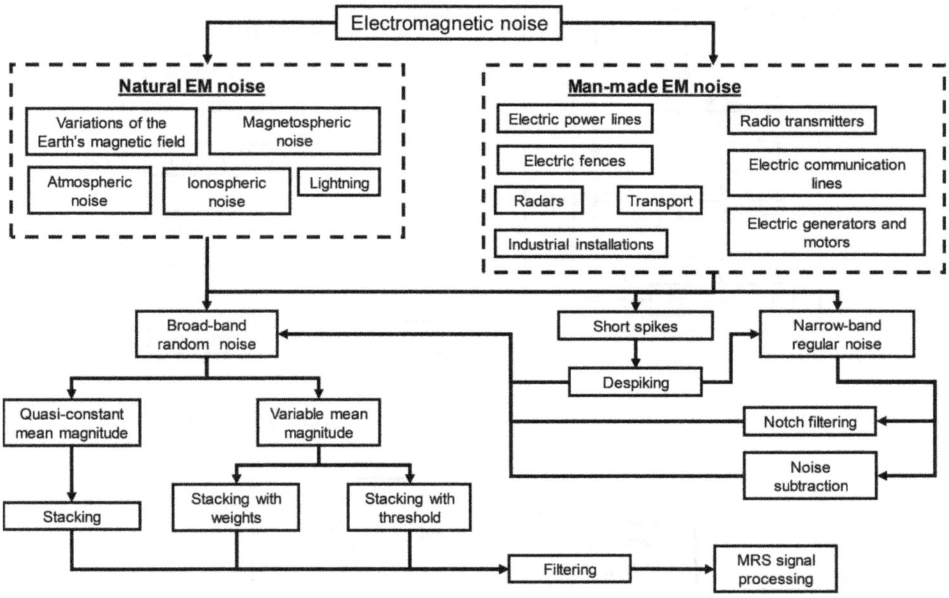


**Figure 3.1.** Ambient electromagnetic noise and corresponding record processing methods.

Power lines carry electricity and the current of the industrial frequency of 50 or 60 Hz contains $N$ harmonics of electrical power supply and random EM noise. The frequency of the $n$th harmonic is $50 \times n$ or $60 \times n$ Hz ($n = 1, 2, \ldots, N$). In the frequency range between 1 and 3 kHz, the noise can represent up to 50% of energy relative to the sinusoidal harmonics. This random component of the EM field transmitted by a power line is generated by a non-ideal sinusoidal current circulating in power lines and by voltage induced by natural EM fields in long wires acting as an antenna that canalizes ambient EM noise. The regular part comprises 50 or 60 Hz harmonics and can be filtered using a notch filter or one of the existing subtraction algorithms. The random component represents a broad-band noise similar to the natural EM noise. A non-correlated noise can be eliminated by a stacking procedure. Stacking can be implemented either directly or using weights for individual records. Records with a noise level higher than an imposed threshold can be rejected as unreliable. Parameters of the MRS signal are computed after filtering.

A one-channel MRS instrument is built using the typical processing scheme presented in figure 3.2. The measuring loop may be coincident Tx/Rx or the transmitting and receiving loops can be separated. The separated loops configuration allows for individual optimization of the Tx and Rx loops (number of turns, thickness of wire, position). The coincident Tx/Rx loop allows for reducing size of MRS gear and simplifies the field implementation. Usually, MRS signal is processed in the time domain after a phase-sensitive detector and a low-pass filter, but the processing schemes implemented in the frequency domain are also reported in the literature.

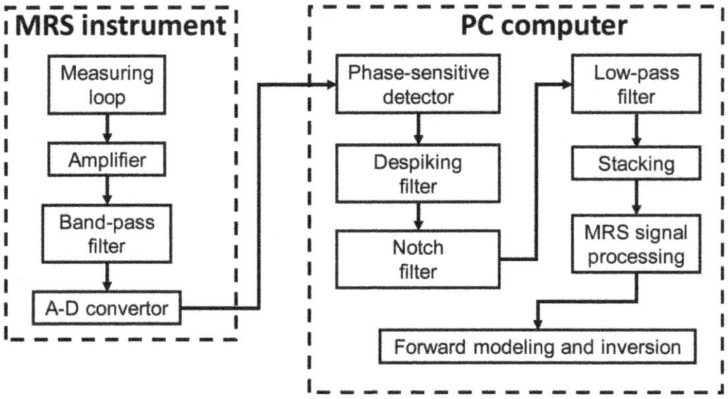


**Figure 3.2.** Typical processing scheme of a one-channel MRS instrument.

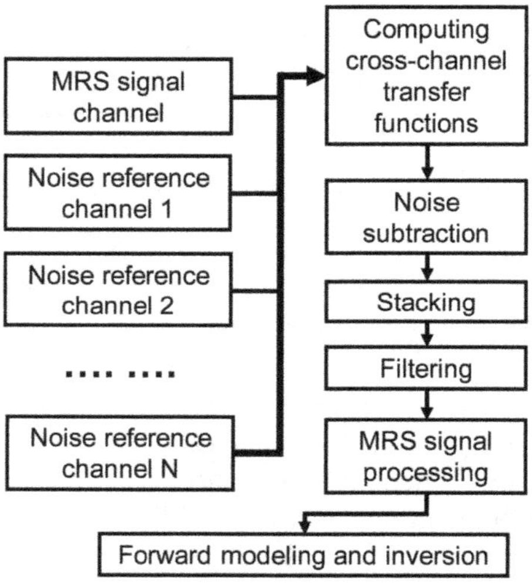


**Figure 3.3.** Typical processing scheme of an MRS instrument with additional noise reference channels.

Each measuring channel of a multi-channel instrument has a similar design to a one-channel instrument. The use of simultaneously recording channels provides two principal advantages: it allows recording noise to not be mixed with the MRS signal and accelerating 2D and 3D prospecting. For applying noise subtraction algorithms, the processing scheme additionally requires computing a cross-channel transfer function (figure 3.3).

Filtering of MRS records will involve taking out noise voltage contaminating the MRS signal. Whatever the filter design will be, it can be analyzed using the frequency spectra of time domain records. The frequency spectra are computed using the Fourier transform. The filter design has two major goals: an efficient elimination of the noise voltage and a preservation of the MRS signal. Figure 3.4 shows frequency spectra of an MRS signal (black line), of a broad-band noise (blue line) and the bandwidth of a filter (red line). Correctly designed filter should have the

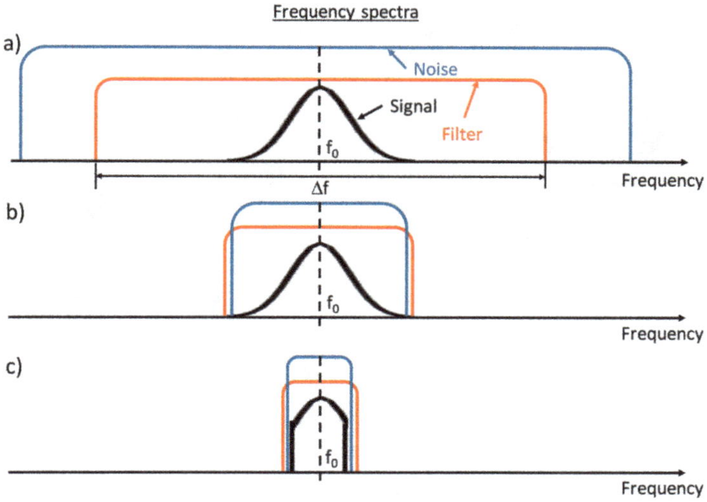


**Figure 3.4.** Frequency spectra of a broad-band noise (blue line), of an MRS signal (black line) and of a band-pass filter (red line): (a) the filter bandwidth ($\Delta f$) is larger than that of the MRS signal and too much noise will be recorded; (b) $\Delta f$ is optimal; (c) $\Delta f$ is narrower than that of the signal and the signal will be disturbed by filtering.

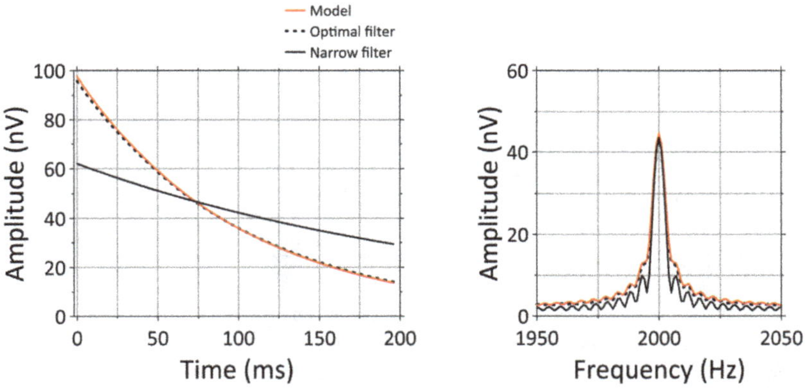


**Figure 3.5.** The left-hand graph shows an exponential signal (red line) filtered using a filter with the optimal bandwidth (black dashed line) and a filter with a too narrow bandwidth (black solid line). The right-hand graph shows the corresponding frequency spectra.

bandwidth shown in figure 3.4(b). Otherwise, the filter either will be inefficient for eliminating the noise (figure 3.4(a)) or will disturb the signal (figure 3.4(c)).

For example, figure 3.5 shows filtering of a synthetic signal (red line) with the optimal filter bandwidth (black dashed line) and with a narrow bandwidth (black solid line). The right-hand graph shows the corresponding frequency spectra.

An example of filtering of this synthetic signal contaminated by a random noise is shown in figure 3.6. After filtering, the signal is still disturbed by noise, but the noise effect is reduced.

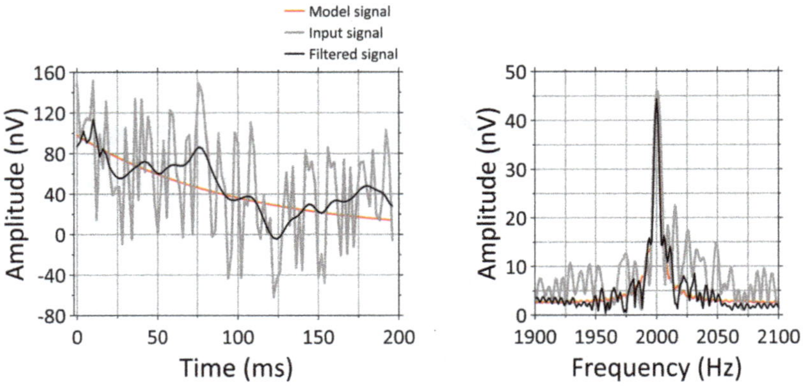


**Figure 3.6.** The left-hand graph shows an exponential signal (red) contaminated with a random noise (gray) and filtered using a filter with the optimal bandwidth (black). The right-hand graph shows the corresponding frequency spectra.

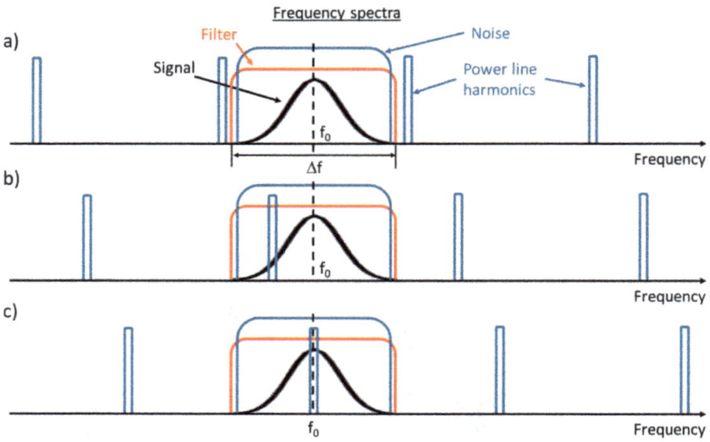


**Figure 3.7.** Frequency spectra of a broad-band noise, of power line harmonics (blue line), of an MRS signal (black line) and of a band-pass filter (red line): (a) the power line harmonics is filtered by the band-pass filter; (b) one of the harmonics is attenuated by the filter; (c) one of the power line harmonics is not attenuated by the filter.

A narrow-band regular noise is produced by electric generators and power lines. An ideal filter can be efficient for filtering power line harmonics when their frequency is outside the band-pass of the filter centered at the Larmor frequency (figure 3.7(a)). Otherwise, the signal will be disturbed (figures 3.7(b), (c)). Unfiltered power line harmonics often have much larger amplitude than the MRS signal and may have a strong effect on the accuracy of MRS measurements.

For example, figure 3.8 shows a synthetic exponential signal ($T_2^* = 100$ ms) centered at 2005 Hz (red line) contaminated by three 50-Hz-harmonics (gray line) and the corresponding frequency spectra.

Filtering of a narrow-band regular noise using a notch filter is based on the representation of the noise by sinusoidal signals with fixed frequencies. The phase

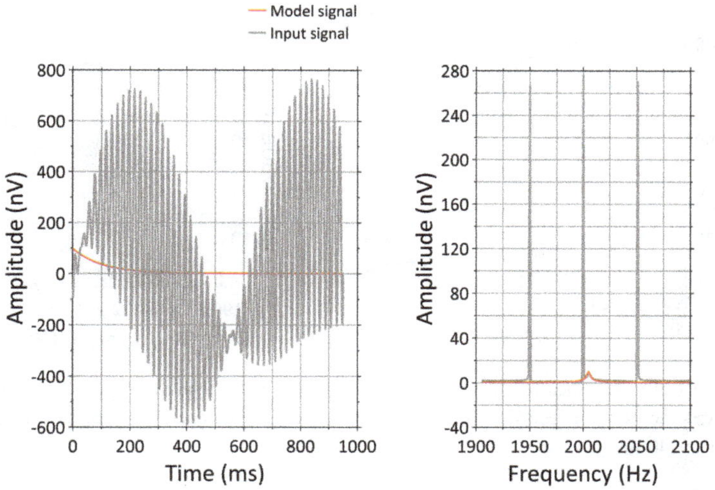


**Figure 3.8.** An exponential signal centered at 2005 Hz (red line) contaminated by three 50-Hz-harmonics (gray line) and the corresponding frequency spectra.

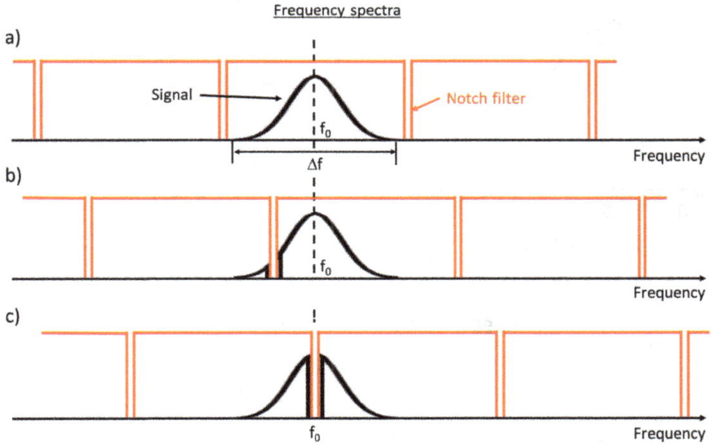


**Figure 3.9.** Effect of the notch filter on the MRS signal for three cases shown in the previous figure: (a) undisturbed MRS signal; (b) slightly disturbed signal; (c) strongly disturbed signal.

and the amplitude of this signal may vary from one record to another. A common example of the regular noise is power line harmonics. If the power line harmonics have frequencies outside the bandwidth of the MRS signal, then the notch filter does not disturb the signal (figure 3.9(a)). If the spectrum of the MRS signal is only partly cut off by the filter (figure 3.9(b)), then the effect of the notch filtering is much less harmful in comparison with the unfiltered power line harmonic. The most difficult case is when the Larmor frequency and one of the power line harmonics frequency are about equal (figure 3.9(c)). In this case, the MRS signal can be strongly disturbed.

Figures 3.10–3.12 show examples of application of the notch filter to an exponential signal centered at 2000, 2010 and 2025 Hz. When the frequency of

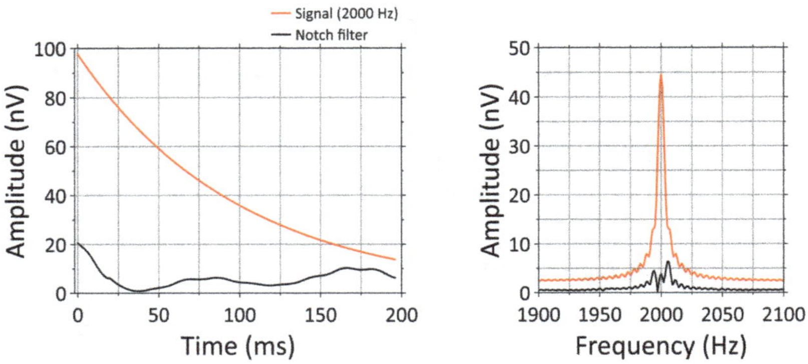


**Figure 3.10.** Effect of the application of a 50-Hz-notch filter to an exponential signal center at 2000 Hz. The left graph shows time domain records and the right graph shows the corresponding frequency spectra. The record length was 240 ms.

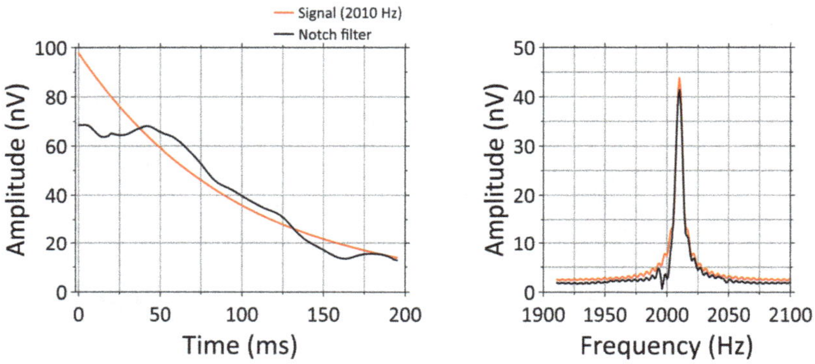


**Figure 3.11.** Effect of the application of a 50-Hz-notch filter to an exponential signal center at 2010 Hz. The left graph shows time domain records and the right graph shows the corresponding frequency spectra. The record length was 240 ms.

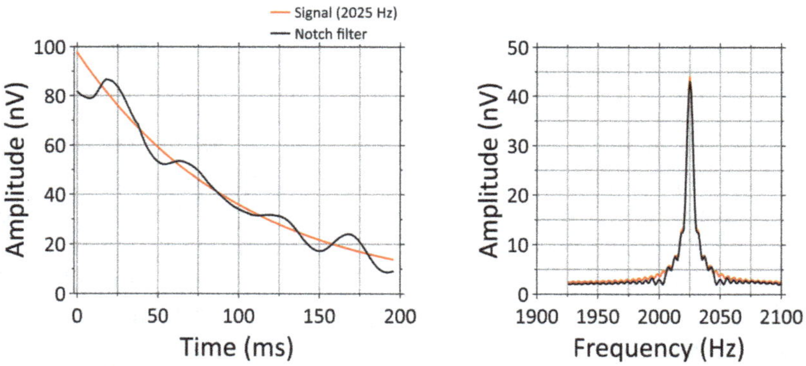


**Figure 3.12.** Effect of the application of a 50-Hz-notch filter to an exponential signal center at 2025 Hz. The left graph shows time domain records and the right graph shows the corresponding frequency spectra. The record length was 240 ms.

the signal and of the closest power line harmonic (2000 Hz) are equal, the signal is almost completely offset by the filter (figure 3.10). But if the separation between the signal and the filter frequencies is larger, then the signal is less disturbed (figures 3.11 and 3.12).

Any filtering is based on differences between the signal and the noise. An easy case is when their spectra are separated. However, it may be not the case when filtering power line harmonics with a notch filter. A notch filter has the bandwidth equal to the bandwidth of the harmonic and may cut off some part of the signal spectrum thus disturbing the signal (figure 3.9). The disturbance can be reduced making the bandwidth of the filter as narrow as possible. The bandwidth of the power line harmonics and consequently of the notch filter can be reduced by making longer records. Figure 3.13 shows spectra of an exponential signal ($T_2^* = 100$ ms) and of a power line harmonic corresponding to the record length of 240 and 1040 ms. When the record length was 240 ms, spectra of the signal and of the harmonic are almost coincident, which explains the strong distortion of the signal observed in figure 3.10. Figure 3.14 shows that the signal spectrum does not change when increasing the record length. Thus, using longer records we expect smaller disturbance of the signal caused by the notch filtering.

An example of the notch filtering of a long record (1040 ms) is shown in figure 3.15. An exponential signal ($T_2^* = 100$ ms) was well recovered after filtering and its spectrum was only slightly distorted. The distortion is smaller than that caused by the filter when the record length was 240 ms (figure 3.11).

Figure 3.16(a) shows an exponential signal ($T_2^* = 100$ ms) centered at 2005 Hz (red line) contaminated by power line harmonics (gray line). The record length was 1040 ms. The result of the notch filtering (figure 3.16(b)) shows that the signal was recovered with a relatively small disturbance.

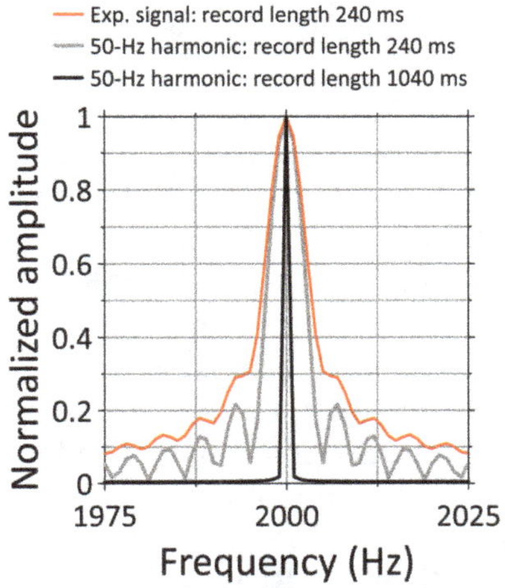


**Figure 3.13.** Frequency spectra of an exponential signal (red) and the bandwidth of the notch filter applied to processing of records with length 240 ms (gray) and 1040 ms (black).

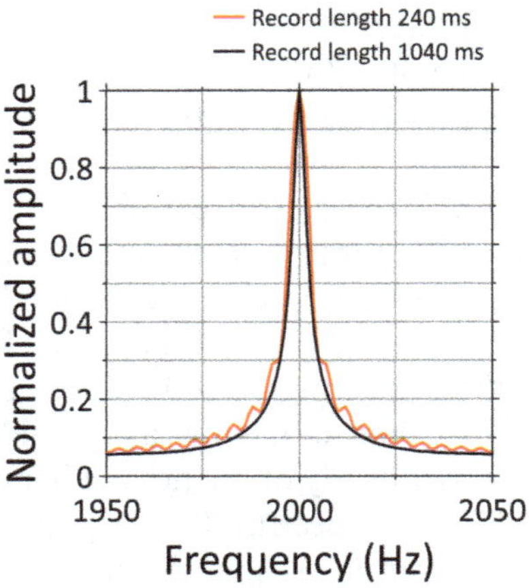


**Figure 3.14.** Frequency spectra of an exponential signal computed considering a 240 ms record length (red) and 1040 ms record length.

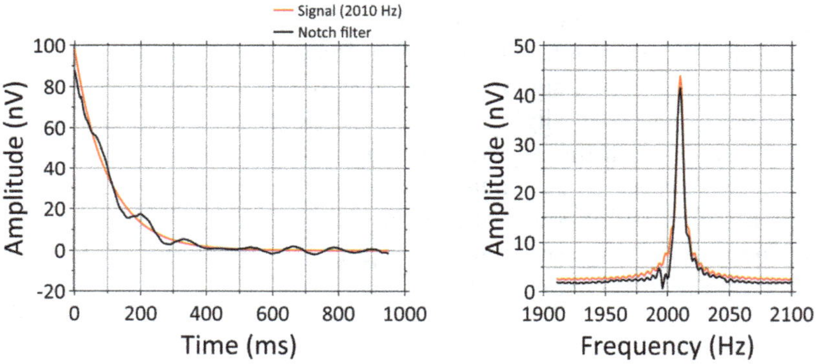


**Figure 3.15.** Effect of the application of a 50-Hz-notch filter to an exponential signal center at 2010 Hz. The left graph shows time domain records and the right graph shows the corresponding frequency spectra. The record length was 1040 ms.

When the difference between the signal frequency and the power line harmonic frequency is less than 5 Hz, the signal will be disturbed by the notch filtering even using long records. In this case, power line harmonics can be filtered using a noise subtraction algorithm that does not disturb the MRS signal. Subtraction of noise can be done using one measuring loop and time-separated records or a few measuring loops and space-separated records.

Figure 3.17(a) shows that for the time-separated subtraction, noise is recorded before injecting the pulse. The noise record should be longer than the signal record. Subtraction is based on two assumptions: noise is regular and hence is the same before and after the pulse; and the noise amplitude is much larger than the signal amplitude. A noise sample (interval B) is selected as similar as possible to the signal with noise sample (interval A) and then, the interval B is subtracted from the interval A.

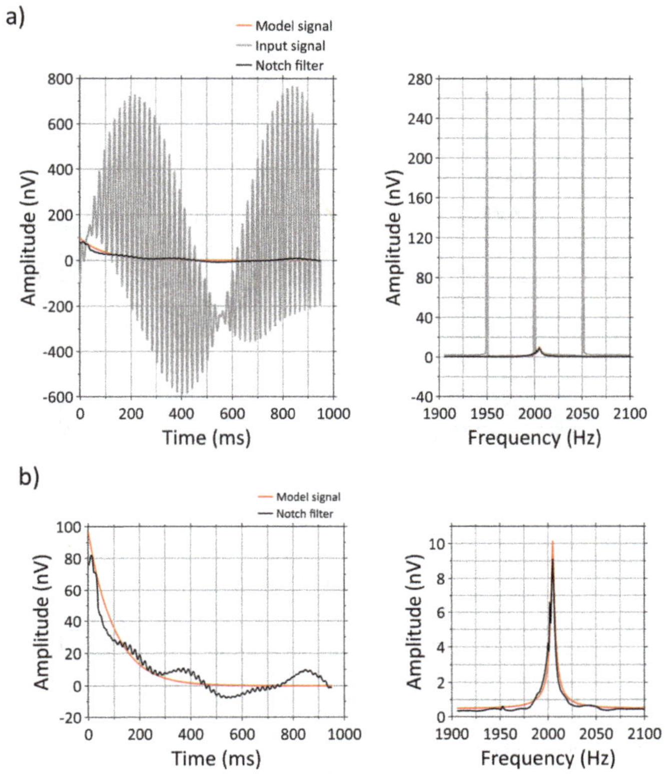


**Figure 3.16.** (a) An exponential signal centered at 2005 Hz (red) and contaminated by three 50-Hz-harmonics (gray) filtered applying a notch signal (black). The record length was 1040 ms. (b) The previous graph shown without the noise-contaminated signal.

Figure 3.17(b) shows that for the space-separated subtraction, simultaneous records are carried out in a few measuring loops. Subtraction is based on the assumption that the same noise patterns are presented in all these records. One loop (MRS signal channel) should be a coincident Tx/Rx loop or the signal receiving Rx loop located close to the Tx loop. Records corresponding to the signal channel contain both the signal and the noise. Other loops (noise channels) are noise reference loops. These loops are set away from the Tx loop so that the MRS signal is not induced in the noise reference loops. Before the pulse, only noise is measured with all the loops. Records before the pulse allow for computing the transfer function between the MRS signal loop and the noise reference loops. This transfer function is used for predicting noise in the signal receiving loop using noise recorded in the noise reference loops. After the pulse, the MRS signal channel measures the MRS signal contaminated by noise and the noise channels measure only noise. Noise predicted in the MRS signal channel is then subtracted from the noise-contaminated signal.

Short spikes represent another type of specific noise. It is characterized by amplitude a few orders of magnitude higher than the MRS signal (figure 3.18(a)).

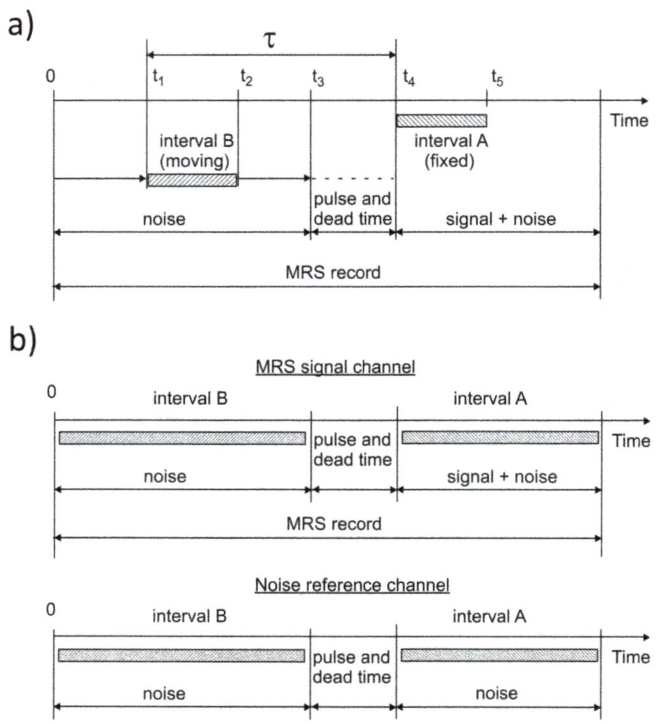


**Figure 3.17.** Noise subtraction methods: (a) time-separated-noise and MRS signal patterns recorded using the same measuring loop; (b) space-separated-noise and MRS signal patterns recorded using separated measuring loops.

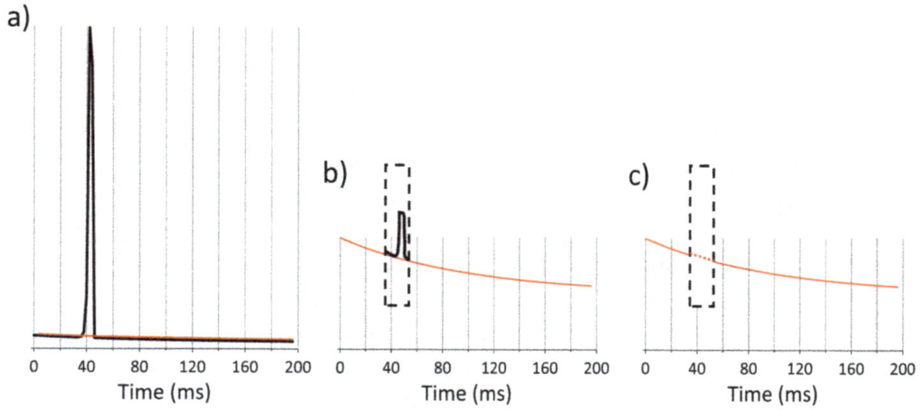


**Figure 3.18.** (a) A short spike disturbing an exponential signal record; (b) identification of the spike location; (c) time domain subtraction of the spike.

The amplitude allows easy identification of spikes before filtering (figure 3.18(b)). Spikes are much shorter than MRS signals and can be cut off in time domain without disturbing the signal (figure 3.18(c)). Different papers report algorithms of spike attenuation in frequency domain.

It is important to eliminate spikes before using a low-pass filter because the amplitude of a short spike can be very high and the frequency spectrum very wide. An example a signal distortion after filtering is shown in figure 3.19.

An efficient method for increasing the signal-to-noise ratio (SNR) is stacking. Stacking comprises averaging up voltage induced in the receiving loop. Each record $\tilde{v}(t)$ contains an MRS signals $v(t)$ contaminated by a broad-band random noise $\varepsilon(t)$ so that $\tilde{v}(t) = v(t) + \varepsilon(t)$. The time-record $\tilde{v}(t)$ is reduced to a discrete-time signal by sampling containing $K$ samples. For each pulse moment, measurements are repeated $N$ times. The data set is a rectangular matrix $\tilde{\mathbf{v}} = [\tilde{v}_{n,k}]$, where $n = 1, 2, \ldots, N$ is the number of stacks and $k = 1, 2, \ldots, K$ is the number of samples. The stacking result is an averaged signal $\bar{\mathbf{v}} = [\bar{v}_k]$, where

$$\bar{v}_k = \frac{1}{N}\sum_{n=1}^{N}\tilde{v}_{k,n} = \frac{1}{N}\sum_{n=1}^{N}(v_{k,n} + \varepsilon_{k,n}). \tag{3.1}$$

MRS signals $v_1(t) = v_2(t) = \ldots = v_n(t) = v(t)$ are the same for each stack $n$. After $N$ stacks, the value of each sample of the MRS signal is $\bar{v}_k = \frac{1}{N}\sum_{n=1}^{N} v_{k,n} = \frac{1}{N}(N \times v_k) = v_k$. Noise however, is not regular and $\varepsilon_1(t) \neq \varepsilon_2(t) \neq \ldots \neq \varepsilon_n(t)$. After $N$ stacks, the value of each sample of a broad-band random noise with a mean $\mu_k = \sqrt{\frac{1}{N}\sum_{n=1}^{N}\varepsilon_{k,n}}$ and a standard deviation $\sigma_k = \sqrt{\frac{1}{N}\sum_{n=1}^{N}\varepsilon_{k,\,n}^{2}}$ converges toward its mean $\bar{\varepsilon}_k = \frac{1}{N}\sum_{n=1}^{N}\varepsilon_{k,n} \to \mu_k$. An approximate convergence rate is $\sqrt{N}$. The stacking result is an averaged signal $\bar{\mathbf{v}} = [\bar{v}_k]$, where

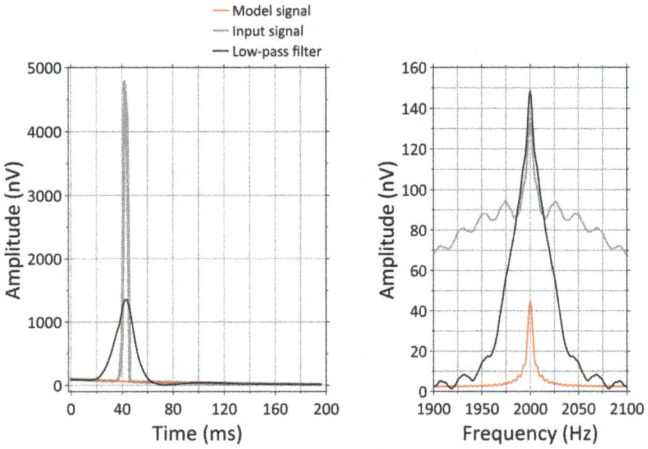


**Figure 3.19.** The left graph shows an exponential signal (red) contaminated by a spike (gray) and filtered applying a low-pass signal with the optimal bandwidth (black). The right graph shows the corresponding frequency spectra.

$$\bar{v}_k = \frac{1}{N}\sum_{n=1}^{N}\tilde{v}_{k,n} = v_{k,n} + \left(\mu_k + \frac{\sigma_k}{\sqrt{N}}\right). \tag{3.2}$$

Thus, stacking preserves the signal and reduces the standard deviation of noise by the factor of $\sqrt{N}$. Natural broad-band noise measured with MRS instrument has often a mean of zero and $\mu_k \to 0$ for $N \to \infty$.

If noise is random, but nonstationary and its magnitude significantly varies from one record to another, then the stacking scheme, namely weighted stacking, can be applied. This method allows maximizing the SNR by giving larger weights to records with smaller noise. For computing weights, we assume that noise is much larger than the MRS signal and voltage recorded after each pulse represents a noise rather than the MRS signal. The weight $s_n$ for the record $n$ is defined as

$$s_n = \frac{1}{\sum_{k=1}^{K}\tilde{v}_{k,n}^2}. \tag{3.3}$$

Then, records are averaged with such weights

$$\bar{v}_k = \frac{\sum_{n=1}^{N}\left(s_n \times \tilde{v}_{k,n}\right)}{\sum_{n=1}^{N}s_n}. \tag{3.4}$$

If the noise magnitude varies a lot, but irregularly, and the time interval between high and low-noise magnitudes is long (more than tens of seconds), then it may be preferable just to skip noisy records. For example, such a noise can be observed during thunderstorms or when measurements are carried out near a road and cars are passing by. Thus, stacking is performed by selecting only low-noise records, which allows the use of a smaller number of stacks. A threshold is set with the noise estimate based either on a standard deviation of the recorded voltage $\sigma_n = \sqrt{\frac{1}{K}\sum_{k=1}^{K}\tilde{v}_{k,n}^2}$ or detecting the maximum voltage value $v_{\max,n} = \max\limits_{k=1,2,..,K}(\tilde{v}_{k,n})$. Both criteria can be used together. The acceptance rule is the following

$$\begin{array}{c}\sigma_n < \sigma_{\lim}\\ \text{or/and}\\ v_{\max,n} < v_{\lim}\end{array} \tag{3.5}$$

where $\sigma_{\lim}$ and $v_{\lim}$ are respectively the thresholds for the standard deviation and the maximum noise voltage amplitude.

## 3.2 Parameters of the MRS signal

An MRS signal is an amplitude modulated (AM) signal. The Larmor frequency is the carrier signal and a modulating signal contains information about amplitude, phase and time constant $T_2^*$ of the MRS signal. The Larmor frequency itself is also one of the parameters of the MRS signal. A popular method for demodulating AM signals is a synchronous detector that recovers information from a modulated signal by mixing the signal with a reference frequency $f_{sd}$ that is a replica of the carrier frequency. Synchronous detection preserves any phase information originally present in the modulating signal. When demodulating MRS signals, $f_{sd}$ is set close to the Larmor frequency $f_0$ ($f_{sd} \approx f_0$). The demodulated signal is presented by two signals in the orthogonal channels $X$ and $Y$. The phase of the reference frequency for the $X$ channel coincides with the phase of the current in the loop, and the phase of the frequency for the $Y$ channel is shifted by an angle of 90 degrees. Channel $X$ shows the real part of an MRS signal $X(t) = \mathrm{Re}(v(t))$ and channel $Y$ shows the imaginary part $Y(t) = \mathrm{Im}(v(t))$. After synchronous detector, a free induction decay (FID) signal is

$$\begin{aligned} X(t) &= v_0 \times \exp(-t/T_2^*) \times \cos(\Delta\omega \times t + \varphi_0) + \varepsilon_X(t) \\ Y(t) &= v_0 \times \exp(-t/T_2^*) \times \sin(\Delta\omega \times t + \varphi_0) + \varepsilon_Y(t) \end{aligned} \tag{3.6}$$

where $\Delta\omega = 2\pi \times \Delta f = 2\pi \times (f_0 - f_{sd})$ and $\varepsilon_X$, $\varepsilon_Y$ is an ambient EM noise.

Neglecting the relaxation and the molecular diffusion, a spin echo signal (SE) with a Gaussian envelope can be written as

$$\begin{aligned} X(t) &= \mathrm{se}_0 \times \exp\left(-\frac{1}{2}\left(\frac{t - 2\tau_d}{T_2^*}\right)^2\right) \times \cos(\Delta\omega \times t + \varphi_0) + \varepsilon_X(t) \\ Y(t) &= \mathrm{se}_0 \times \exp\left(-\frac{1}{2}\left(\frac{t - 2\tau_d}{T_2^*}\right)^2\right) \times \sin(\Delta\omega \times t + \varphi_0) + \varepsilon_Y(t). \end{aligned} \tag{3.7}$$

Equations (3.6 and 3.7) show that the phase measured after the synchronous detector is independent of the amplitude

$$\Phi(t) = \tan^{-1}\left(\frac{Y(t)}{X(t)}\right) = \Delta\omega \times t + \varphi_0. \tag{3.8}$$

It is an equation of the straight line shifted at the initial phase $\varphi_0$ and inclined at an angle $\Delta\omega$. For computing $\Delta\omega$ and $\varphi_0$, measured phase $\Phi(t)$ is fitted by a straight line. For that, a normal matrix equation should be sold

$$\mathbf{A}^{\mathrm{T}}\mathbf{A}\mathbf{x} = \mathbf{A}^{\mathrm{T}}\mathbf{b} \tag{3.9}$$

where $\mathbf{A} = \begin{pmatrix} \Delta t \times 1 & 1 \\ \Delta t \times 2 & 1 \\ \cdots & 1 \\ \Delta t \times k & 1 \\ \cdots & 1 \\ \Delta t \times K & 1 \end{pmatrix}$ is a rectangular matrix, $\mathbf{b} = \begin{pmatrix} v_1 \\ v_2 \\ \cdots \\ v_k \\ \cdots \\ v_K \end{pmatrix}$ is a vector containing signal samples with $k = 1, 2, \ldots, K$ being the number of samples, $\mathbf{A}^T$ is the transpose of a matrix $\mathbf{A}$, $\mathbf{x} = [\Delta\omega \quad \varphi_0]$ is a solution vector.

The amplitude of the FID and SE signals is independent of $\Delta\omega$ and $\varphi_0$, but the shapes of these signals are different. Equation (3.10) shows the amplitude of FID signal

$$v(t) = \sqrt{X^2(t) + Y^2(t)} = v_0 \times \exp(-t/T_2^*). \tag{3.10}$$

For solving equation (3.10), it is convenient to use the natural logarithm

$$\log_e(v(t)) = \log_e(v_0) - t/T_2^*. \tag{3.11}$$

Assuming $\alpha = -1/T_2^*$ and $\beta = \log_e(v_0)$, equation (3.11) can be written as

$$\log_e(v(t)) = \alpha t + \beta. \tag{3.12}$$

It is an equation of the straight line. In matrix notation, it is a normal matrix equation (equation 3.9), where $\mathbf{b} = \begin{pmatrix} \log_e v_1 \\ \log_e v_2 \\ \cdots \\ \log_e v_k \\ \cdots \\ \log_e v_K \end{pmatrix}$ is a vector containing signal samples, $\mathbf{x} = [\alpha \quad \beta]$ is a solution vector. Then,

$$\begin{aligned} T_2^* &= -1/\alpha \\ v_0 &= \exp(\beta). \end{aligned} \tag{3.13}$$

Figure 3.20 shows a synthetic noiseless signal computed assuming $v_0 = 100$ nV. $T_2^* = 200$ ms, $\varphi_0 = -85°$ and $\Delta f = -2$ Hz. This example shows that the phase and the logarithm of the amplitude are linear functions of time.

If the signal-to-noise ratio (SNR) is high, then the linear regression fit presented above provides a reliable estimate of the signal parameters. In practice, a more stable non-linear regression curve-fitting technique based on the least squares minimization is often used. For using this technique, measured after synchronous detector signals $X(t)$, $Y(t)$ are fitted by a synthetic signal $\tilde{X}(t)$, $\tilde{Y}(t)$ computed as

$$\begin{aligned} \tilde{X}(t) &= \tilde{v}_0 \times \exp(-t/\tilde{T}_2^*) \times \cos(\Delta\tilde{\omega} \times t + \tilde{\varphi}_0) \\ \tilde{Y}(t) &= \tilde{v}_0 \times \exp(-t/\tilde{T}_2^*) \times \sin(\Delta\tilde{\omega} \times t + \tilde{\varphi}_0) \end{aligned} \tag{3.14}$$

where $\tilde{v}_0$, $\tilde{T}_2^*$, $\tilde{\varphi}_0$, $\Delta\tilde{\omega}$ are parameters of the synthetic signal minimizing the following equation

$$\frac{1}{2K}\sum_{k=1}^{K}\left((x_k - \tilde{x}_k)^2 + (y_k - \tilde{y}_k)^2\right) < \sigma_\varepsilon^2 \tag{3.15}$$

where $x_k$, $y_k$ are measured samples, $\tilde{x}_k$, $\tilde{y}_k$ are synthetic samples ($k = 1, 2, \ldots, K$), $\sigma_\varepsilon$ is a standard deviation of noise. If $\sigma_\varepsilon \to 0$, then $v_0 \cong \tilde{v}_0$, $T_2^* \cong \tilde{T}_2^*$, $\varphi_0 \cong \tilde{\varphi}_0$, $\Delta\omega \cong \Delta\tilde{\omega}$. Any suitable optimization method can be used for the minimization. For example, the Marquardt non-linear fitting algorithm.

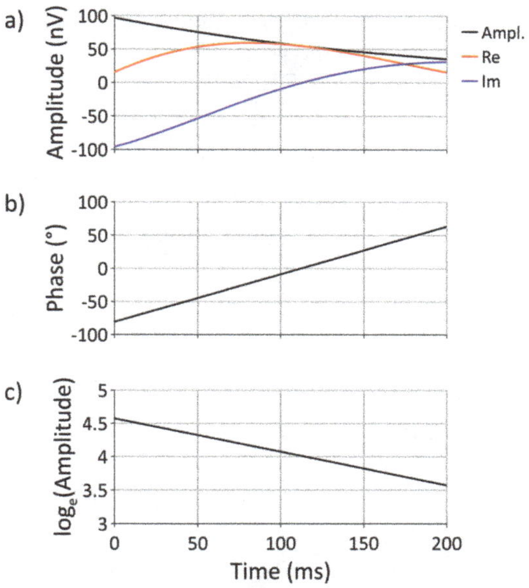


**Figure 3.20.** Example of a synthetic signal ($v_0 = 100$ nV, $T_2{}^* = 200$ ms, $\varphi_0 = -85°$, $\Delta f = -2$ Hz). (a) The signal voltage versus time. (b) The phase versus time. (c) The natural logarithm of the amplitude versus time.

In laboratory NMR and NMR logging, a multi-exponential $T_2$ inversion of NMR relaxation signals is used for a pore structure characterization. This approach also applies to MRS measurements using an $T_2^*$ inversion. For that, the model for inversion is composed of several synthetic signals

$$\tilde{X}(t) = \sum_{m=1}^{M}\left(\tilde{v}_{0,m} \times \exp(-t/\tilde{T}^*_{2,\,m}) \times \cos(\Delta\tilde{\omega}_m \times t + \tilde{\varphi}_{0,m})\right)$$
$$\tilde{Y}(t) = \sum_{m=1}^{M}\left(\tilde{v}_{0,m} \times \exp(-t/\tilde{T}^*_{2,\,m}) \times \sin(\Delta\tilde{\omega}_m \times t + \tilde{\varphi}_{0,m})\right) \tag{3.16}$$

where $m = 1, 2, \ldots, M$ is a number of signals in the model. For simplicity, it is assumed that for any $m$, the Larmor frequency $\Delta\omega_m = \Delta\omega$ and the phase $\varphi_{0,m} = \varphi_0$ are fixed and only amplitude is used for the multi-exponential inversion

$$\tilde{v}(t) = \sum_{m=1}^{M}\left(\tilde{v}_{0,m} \times \exp(-t/\tilde{T}^*_{2,\,m})\right). \tag{3.17}$$

The multi-exponential inversion for $\tilde{v}_{0,m}$ and $\tilde{T}^*_{2,\,m}$ comprises the least-square minimization

$$\frac{1}{K}\sum_{k=1}^{K}(v_k - \tilde{v}_k)^2 < \sigma_\varepsilon^2. \tag{3.18}$$

Let us consider the following numerical example. Table 3.1 shows a model composed of three-exponential signals (initial model). A noiseless synthetic signal

**Table 3.1.** A noiseless initial model composed of three-exponential signals with different time constants fitted by models composed of three-, two- and one-exponential signals.

| Model | $v_{0,1}$ (nV) | $T^*_{2,1}$ (ms) | $v_{0,2}$ (nV) | $T^*_{2,2}$ (ms) | $v_{0,3}$ (nV) | $T^*_{2,3}$ (ms) | RMSE (nV) |
|---|---|---|---|---|---|---|---|
| Initial model | 100 | 40 | 100 | 100 | 100 | 250 | |
| Fit: 3-exponential | 99 | 40 | 99.5 | 99 | 101 | 249 | 0.0009 |
| Fit: 2-exponential | 138 | 46 | 162 | 193 | | | 0.1 |
| Fit: 1-exponential | 274 | 120 | | | | | 8.3 |

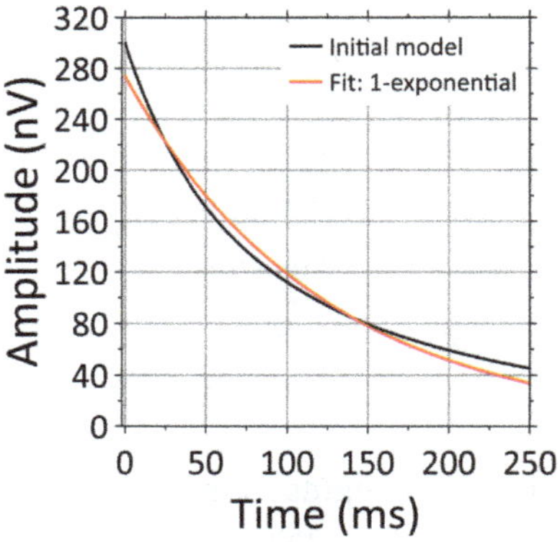


**Figure 3.21.** A signal computed using the three-exponents model fitted by a signal computed one-exponent model with the RMSE = 8.3 nV.

computed considering this model was fitted using models with $M = 3$, $M = 2$ and $M = 1$. $\tilde{v}_{0,m}$ and $\tilde{T}^*_{2,m}$ corresponding to these models as well as the root-mean-square error (*RMSE*) of the fit are also shown in table 3.1.

The models with 3 and 2 exponents fit the initial signal with a high accuracy. The fitting error obtained using the one-exponential model is much larger than that obtained using the multi-exponential models. Figure 3.21 shows the fit with the one-exponential model. The misfit between the noiseless signals is clearly visible, but it is comparable with experimental errors typical for MRS data. A rule of thumb for resolving two-exponential signals is to have very accurate measurements and processing signals with significantly different relaxation times. Three-exponential signals cannot be reliably resolved. It is one of the reasons why the multi-exponential inversion is rarely used for processing MRS measurements.

A similar approach can be used for processing spin echo signals. When using SE, the relaxation time $T_2^*$, the phase $\varphi_0$ and the Larmor frequency $\omega_0$ are strongly disturbed by rocks and have a little hydrological meaning. The SE amplitude and the relaxation time $T_2$ are the parameters of interest. Measurements of $T_2$ are based on the amplitudes of individual echoes as have been discussed in the previous chapters, but knowledge of $T_2^*$ is necessary for computing the amplitude $se_0$ being the principal target. Measured SE amplitude is

$$\text{se}(t) = \text{se}_0 \times \exp\left(-\frac{1}{2}\left(\frac{t - \tau_d}{T_2^*}\right)^2\right) \tag{3.19}$$

where se($t$) is the measured SE signal, $\tau_d$ is the echo arrival time after the second pulse. The natural logarithm of the amplitude yields

$$\log_e(\text{se}(t)) = \log_e(\text{se}_0) - \frac{1}{2(T_2^*)^2}(t - \tau_d)^2. \tag{3.20}$$

Assuming $\alpha = -\frac{1}{2(T_2^*)^2}$ and $\beta = \log_e(\text{se}_0)$, equation (3.20) can be written as

$$\log_e(\text{se}(t)) = \alpha(t - \tau_d)^2 + \beta. \tag{3.21}$$

In matrix notation

$$\mathbf{Ax} = \mathbf{b} \tag{3.22}$$

where $\mathbf{A} = \begin{pmatrix} (\Delta t \times 1 - \tau_d)^2 & 1 \\ (\Delta t \times 2 - \tau_d)^2 & 1 \\ \cdots & 1 \\ (\Delta t \times k - \tau_d)^2 & 1 \\ \cdots & 1 \\ (\Delta t \times K - \tau_d)^2 & 1 \end{pmatrix}$ is a rectangular matrix, where $\mathbf{b} = \begin{pmatrix} \log_e(\text{se}_1) \\ \log_e(\text{se}_2) \\ \cdots \\ \log_e(\text{se}_k) \\ \cdots \\ \log_e(\text{se}_K) \end{pmatrix}$ is a vector containing signal samples, $\mathbf{x} = [\alpha \quad \beta]$ is a solution vector.

Equation (3.22) can be solved for $\alpha$ and $\beta$ as a normal matrix equation (equation 3.9). Then,

$$\begin{aligned} T_2^* &= \sqrt{-1/2\alpha} \\ \text{se}_0 &= \exp(\beta). \end{aligned} \tag{3.23}$$

To avoid computing the logarithm of noisy data, equation (3.19) can be solved using a non-linear regression curve-fitting technique based on the least squares minimization. For that, $\alpha$ and $\beta$ are selected so that a synthetic SE signal $\tilde{\text{se}}(t)$

$$\tilde{\text{se}}(t) = \exp(\alpha(t - \tau_d)^2 + \beta) \tag{3.24}$$

fits a measured SE signal se($t$) with the accuracy defined by an experimental error $\sigma_\varepsilon$

$$\frac{1}{K}\sum_{k=1}^{K}(\text{se}_k - \tilde{\text{se}}_k)^2 < \sigma_\varepsilon^2. \tag{3.25}$$

For the optimization, the Marquardt non-linear fitting algorithm can be used.

**IOP** Publishing

Surface NMR for Hydrogeology

A user's guide

**Anatoly Legchenko**

# Chapter 4

# MRS forward modeling

Forward numerical modeling of magnetic resonance sounding (MRS) data comprises the use of geological models of the Earth and a set of theoretical equations for the MRS response to simulate field experiments with the MRS method. Models can be one, two, or three dimensional and consist of subsurface structures with the associated water contents and relaxation times of the MRS signal. Forward modeling is also used for interpretation of measured data in complex geological environments. The more complex the geometry, the more factors can be modeled but the slower the computing time. Forward modeling is also a useful tool to develop physical understanding in an educational setting. The forward modeling procedure is illustrated in figure 4.1.

A more sophisticated forward modeling applies a sequential procedure. The sequential procedure assumes the separation between flow model results provided by a reservoir simulator and MRS forward modeling results. The workflow steps can be performed as a sequence of independent workflow actions using separate software packages, or they can be performed in a software package designed to simplify the data transfer process. The sequential workflow process can be applied only if the properties used in the flow model are also used in the forward modeling program. Otherwise, the data sets are inconsistent. The integrated flow model is a more efficient tool for integrating flow and petrophysical calculations because it makes the forward modeling step a subroutine of the flow simulator, and data consistency is guaranteed.

## 4.1 Modeling procedure

MRS forward modeling comprises computing the FID signal $v_0(q)$ using equation (2.19) or using equation (2.23) for computing the spin echo (SE) signal $se_0(q)$. Modeling can be carried out assuming 1D, 2D or 3D geometry of water-bearing formations. Considering the 1D case, we assume the horizontal stratification and equation (2.19) can be written as

doi:10.1088/978-0-7503-3155-5ch4  

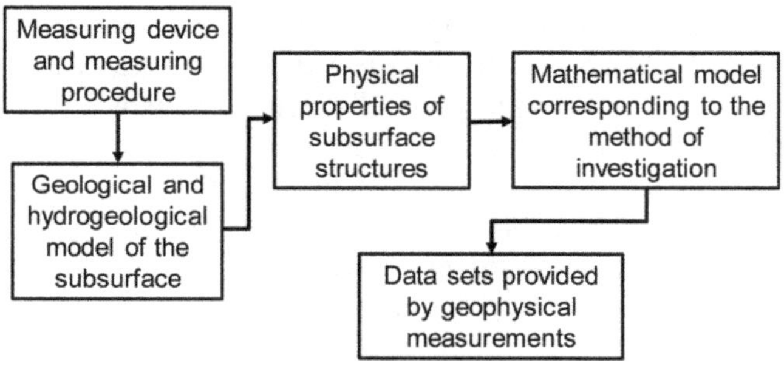


**Figure 4.1.** A flowchart of the forward modeling procedure.

$$v_0(q) = \int_{z=0}^{z=\infty} K_{1D,\,\mathrm{FID}}(q,\, z)w(z)dz \tag{4.1}$$

where $w(z)$ is the solution function and the kernel (also termed the linear filter) $K_{1D,\,\mathrm{FID}}(q,\, z)$ a known function of two variables

$$K_{1D,\,\mathrm{FID}}(q,\, z) = I_0^{-1}M_0\omega_0 \int_{y=-\infty}^{y=\infty}\int_{x=-\infty}^{x=\infty} e^{j\phi_0(\mathbf{r})}B_1 \sin(\theta_p)dxdy. \tag{4.2}$$

The SE signal can be computed as

$$\mathrm{se}_0(q) = \int_{z=0}^{z=\infty} K_{1D,\mathrm{SE}}(q,\, z)w(z)dz \tag{4.3}$$

where

$$K_{1D,\mathrm{SE}}(q,\, z) = I_0^{-1}M_0\omega_0 \int_{y=-\infty}^{y=\infty}\int_{x=-\infty}^{x=\infty} e^{j\phi 0(\mathbf{r})}B_1 \sin^3(\theta_p)dxdy. \tag{4.4}$$

The similarity of the FID and the SE kernels allows for presenting the 2D and 3D cases referring only to the FID signal. In the 2D case, the FID signal is

$$v_0(q) = \int_{z=0}^{z=\infty}\int_{x=-\infty}^{x=\infty} K_{2D,\,\mathrm{FID}}(q,\, x,\, z)w(x,\, z)dxdz \tag{4.5}$$

where $w(x,\, z)$ is a 2D distribution of the water content and the 2D kernel is

$$K_{2D,\,\mathrm{FID}}(q,\, x,\, z) = I_0^{-1}M_0\omega_0 \int_{y=-\infty}^{y=\infty} e^{j\phi 0(\mathbf{r})}B^1 \sin(\theta_p)dy. \tag{4.6}$$

The FID signal generated by a 3D structure is

$$v_0(q) = \int_{z=0}^{z=\infty}\int_{y=-\infty}^{y=\infty}\int_{x=-\infty}^{x=\infty} K_{3D,\,\mathrm{FID}}(q,\, x,\, y,\, z)w(x,\, y,\, z)dxdydz \tag{4.7}$$

with the kernel

$$K_{3D,\,\mathrm{FID}}(q,\, x,\, y,\, z) = I_0^{-1}M_0\omega_0 e^{j\phi 0(\mathbf{r})}B_1 \sin(\theta_p). \tag{4.8}$$

For performing numerical modeling, the integral equation has to be discretized. Discretization comprises defining the number and the values of the pulse moments $\mathbf{q} = [q_1, q_2, .., q_i, .., q_I]$ and representation of the subsurface by a set of homogeneous volumes (cells). The number and the dimension of these cells depend on the complexity of the subsurface geometry. Figure 4.2(a) shows a 1D discretization. The subsurface below the measuring loop is represented by a set of infinite horizontal layers with the water content and the relaxation times invariable within each layer. However, the water content and the relaxation times may vary from one layer to another. Figure 4.2(b) shows a 2D discretization that comprises a representation by a set of rectangular boxes infinite in one direction ($X$ or $Y$). The water content and the relaxation times are assumed invariable within each box, but may vary from one box to another. Figure 4.2(c) shows a 3D discretization. The subsurface is represented by a set of finite homogeneous boxes. Note that the size and the shape of the cells may vary and are usually determined by taking into account each configuration of modeling.

After discretization, equation (4.1) can be approximated by a matrix equation

$$\mathbf{Aw} = \mathbf{v}0 \tag{4.9}$$

where $\mathbf{A} = [a_{i,j}]$ is a rectangular matrix representing the kernel with $i = 1, 2, .., i, .., I$ the number of pulse moments and $j = 1, 2, .., j, .., J$ the number of cells, $\mathbf{w} = [w_j]$ is the water content and $\mathbf{v}0 = [v0_i]$ is the initial amplitude of the FID signal. Considering a 1D discretization, the elements of the matrix $\mathbf{A}$ for the FID signal are computed as

$$a_{i,j} = \left(I_0^{-1}\right)_i M_0\omega_0 \int_{z_j}^{z_j+\Delta z_j} \int_{y=-\infty}^{y=\infty} \int_{x=-\infty}^{x=\infty} e^{j\phi_0(\mathbf{r})}(B_1)_i \sin((\theta_p)_i)dxdydz. \tag{4.10}$$

Equation (4.3) can be approximated by a matrix equation

$$\mathbf{Aw} = \mathbf{se}0 \tag{4.11}$$

where se0 = [se0$i$] is the initial amplitude of the SE signal and the elements of the matrix $\mathbf{A}$ for the SE signal are computed as

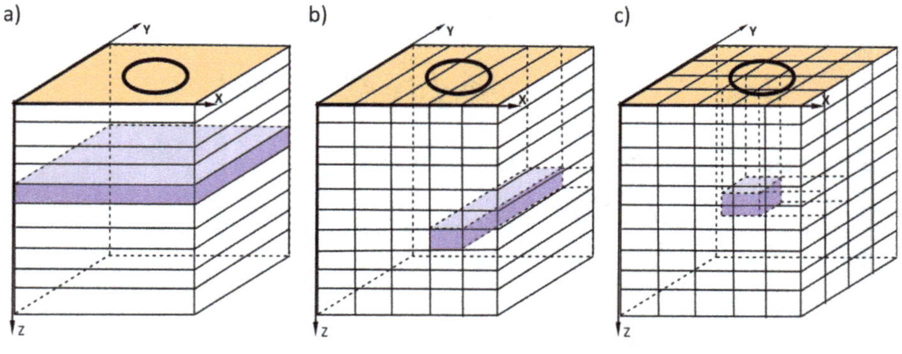


**Figure 4.2.** Geometrical representation of the kernel of MRS integral equation: (a) 1D kernel; (b) 2D kernel; (c) 3D kernel.

$$a_{i,j} = \left(I_0^{-1}\right)_i M_0\omega_0 \int_{z_j}^{z_j+\Delta z_j} \int_{y=-\infty}^{y=\infty} \int_{x=-\infty}^{x=\infty} e^{j\phi_0(\mathbf{r})}(B_1)_i \sin^3((\theta_p)_i)dxdydz. \quad (4.12)$$

Considering the 2D and 3D discretization, the elements of the matrix **A** for the FID signal can be respectively computed as

$$a_{i,j} = \left(I_0^{-1}\right)_i M_0\omega_0 \int_{z_j}^{z_j+\Delta z_j} \int_{x_j}^{x_j+\Delta x_j} \int_{y=-\infty}^{y=\infty} e^{j\phi_0(\mathbf{r})}(B_1)_i \sin((\theta_p)_i)dydxdz \quad (4.13)$$

and

$$a_{i,j} = \left(I_0^{-1}\right)_i M_0\omega_0 \int_{z_j}^{z_j+\Delta z_j} \int_{y_j}^{y_j+\Delta x_j} \int_{x_j}^{x_j+\Delta x_j} e^{j\phi_0(\mathbf{r})}(B_1)_i \sin((\theta_p)_i)dxdydz. \quad (4.14)$$

Taking into account relaxation of the FID signal during the dead time $\Delta t_{\text{dead}}$, equation (4.9) becomes

$$\mathbf{ATw} = \mathbf{v1} \quad (4.15)$$

where a diagonal matrix **T**

$$\mathbf{T} = \begin{bmatrix} RF_1 & 0 & 0 & \dots & 0 & 0 \\ 0 & RF_2 & 0 & \dots & 0 & 0 \\ 0 & 0 & \dots & 0 & 0 & 0 \\ \dots & \dots & 0 & RF_j & 0 & 0 \\ 0 & 0 & \dots & 0 & \dots & 0 \\ 0 & 0 & \dots & 0 & 0 & RF_J \end{bmatrix} \quad (4.16)$$

and the relaxation factor is

$$RF_j = \exp\left(-\Delta t_{\text{dead}}/(T_2^*)_j\right). \quad (4.17)$$

The FID2 signal after the second pulse is

$$\mathbf{ATw} = \mathbf{v2} \quad (4.18)$$

with the relaxation factor computed considering the dead time $\Delta t_{\text{dead}}$ and the delay between the pulses $\tau_d$

$$RF_j = \exp\left(-\Delta t_{\text{dead}}/(T_2^*)_j\right) \times \left(1 - \exp\left(-\tau_d/(T_1)_j\right)\right). \quad (4.19)$$

The SE signal (equation (4.11)) attenuated by the relaxation can be computed as

$$\mathbf{ATw} = \mathbf{se\tau} \quad (4.20)$$

where $\tau_d$ is the delay between the pulses and

$$RF_j = \exp(-2\tau_d/(T_2)_j). \quad (4.21)$$

A real hydrogeological system is composed of different geological formations, some of them representing aquifers. These formations can be rather heterogeneous and their geometry and heterogeneity are known only approximately. The study area may be in any place around the world with different geophysical and meteorological conditions (figure 4.3).

For MRS modeling, a real system should be approximated by simplified geological formations with their physical properties (figure 4.4). Often, the subsurface is represented by horizontal or inclined homogeneous layers of thickness $\Delta z$ at the depth $z$. The electrical resistivity $\rho$ and the Larmor frequency $f_0$ should be known or assumed. If the geometry of the subsurface is known with a high precision, then the geological model can be more complex.

Water in the subsurface is approximated by a set of water-bearing formations of simplified geometry: horizontal layers in a 1D case or by boxes in a 2D or 3D case. For each formation, the depth $z$, the thickness $\Delta z$, the water content $w$, the relaxation times $T_2^*$ and either $T_1$ or $T_2$ should be known or assumed.

Figure 4.5 shows typical loops applied with the MRS. Loops may be circular or square. The figure-eight loops allow for a better signal-to-noise ratio and these loops are often used in a noisy environment. The size of the loop may vary between 25-m-side and 150-m-side for a square loop and 25 and 150 m in diameter for a circular loop. To be accorded to the current generator of the MRS instrument, the inductance of the transmitting loop should be between 0.5 and 1.5 mH. For that,

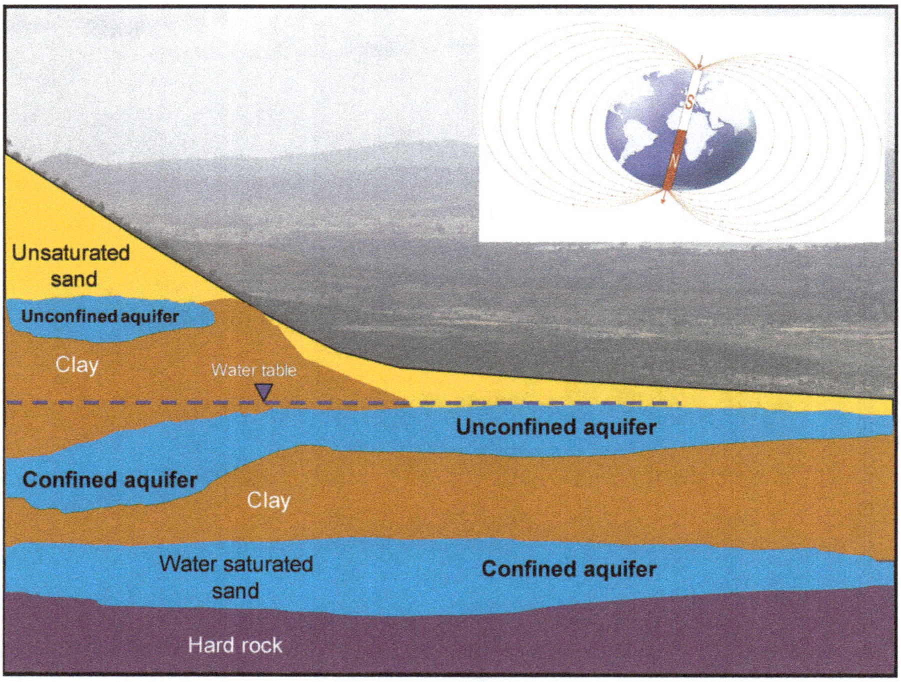


**Figure 4.3.** Example of a hydrogeological system to study.

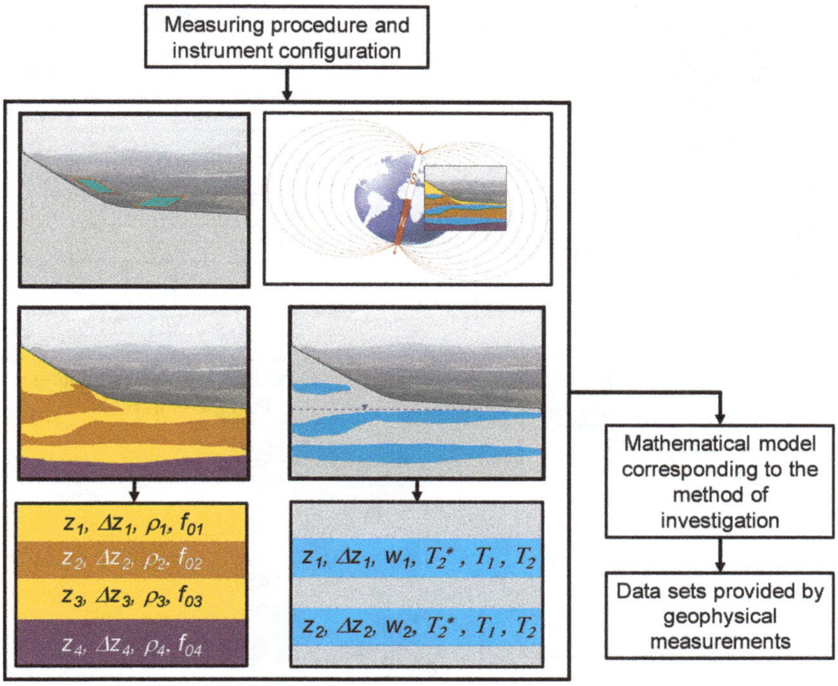


**Figure 4.4.** Numerical representation of the hydrogeological system for forward modeling with the MRS method.

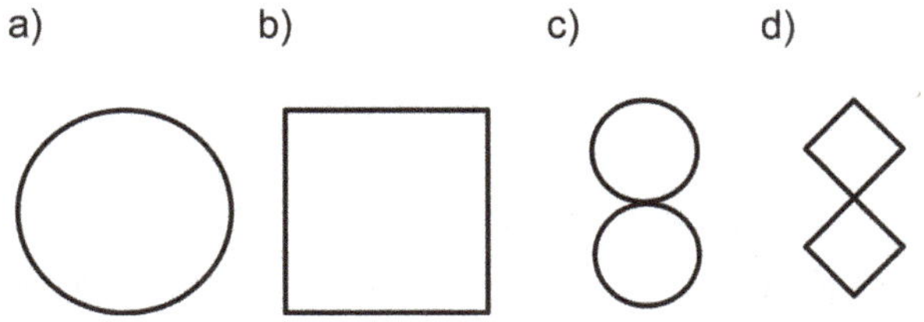


**Figure 4.5.** Typical MRS loops: (a) circular; (b) square; (c) circular figure-eight; (d) square figure-eight.

small loops (<50-m-side/diameter for a square or a circular loop; or <30-m-side/diameter for a figure-eight loop) should have two turns of wire. With the same wire length, circular loops (figures 4.5(a), (b)) have larger surface area than the square loops. For example, with a 314-m-long wire, a circular loop has the surface area of 7850 $m^2$. A square loop made of the same wire has the surface area of 6162,25 $m^2$, which is 21.5% less than the circular loop. MRS loops are made of a thick wire with a cross-section of at least 6 $mm^2$. This wire is heavy and expensive and a more efficient use of the wire is an advantage of circular loops. In practice however, square loops are much easier to install and, for this reason, circular loops are less popular.

Figure 4.6 summarizes the main steps to follow for performing MRS forward modeling.

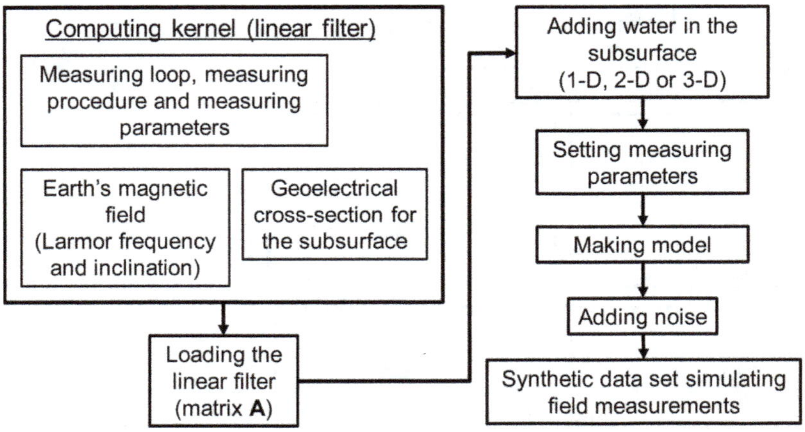


**Figure 4.6.** MRS forward modeling procedure.

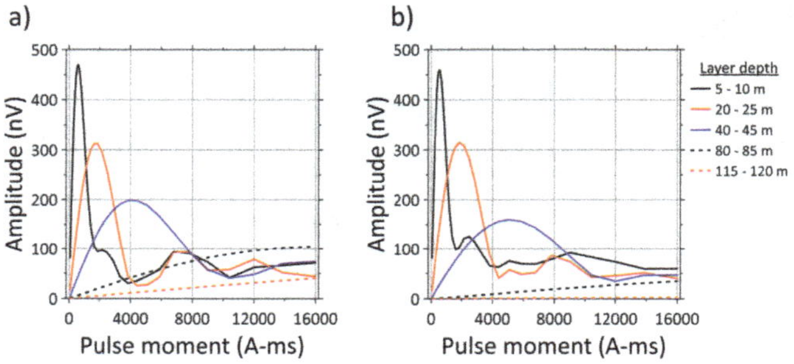


**Figure 4.7.** Amplitude of the MRS signal versus pulse moment computed considering different depths of a 5-m-thick water-saturated layer: (a) 100-Ω.m-half-space; (b) 10-Ω.m-half-space.

## 4.2 Examples of the MRS response in layered subsurface

For using the MRS method, it is important to get a glimpse of what sort of measured signals one will get in typical measuring conditions. For modeling, we assume a 5-m-thick horizontal layer with the water content of $w = 0.4$ and the relaxation time $T_2^* = 400$ ms; the Earth's magnetic field of 46 938 nT with the corresponding Larmor frequency of 2000 Hz; the inclination of 55°N; and a circular loop of 100 m in diameter.

Figure 4.7 shows the amplitude of the MRS signal versus pulse moment ($v1(q)$) computed assuming a homogeneous subsurface with the resistivity of 100 Ω.m (figures 4.7(a)) and 10 Ω.m (figure 4.7(b)). Results show that the amplitude $v1(q)$ has a pronounced maximum and a varying shape, both depending on the layer depth.

The maximum of the amplitude plotted versus depth is shown in figure 4.8(a) and the value of the pulse moment versus depth corresponding to the maximum amplitude is shown in figure 4.8(b). Figure 4.8 shows that in a 10-Ω.m-half-space in comparison to a 100-Ω.m-half-space, the MRS response is attenuated and a larger pulse moment is necessary for getting the maximum signal.

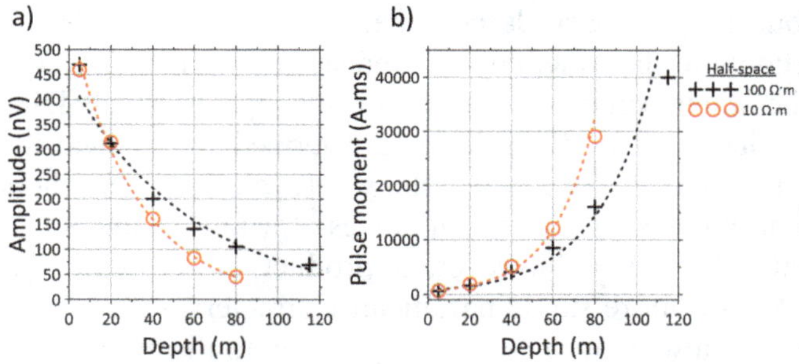


**Figure 4.8.** The maximum amplitude of the MRS signal generated by a 5-m-thick water-saturated layer versus layer depth: (a) 100-Ω.m-half-space; (b) 10-Ω.m-half-space. Dashed lines show exponential fits of these data.

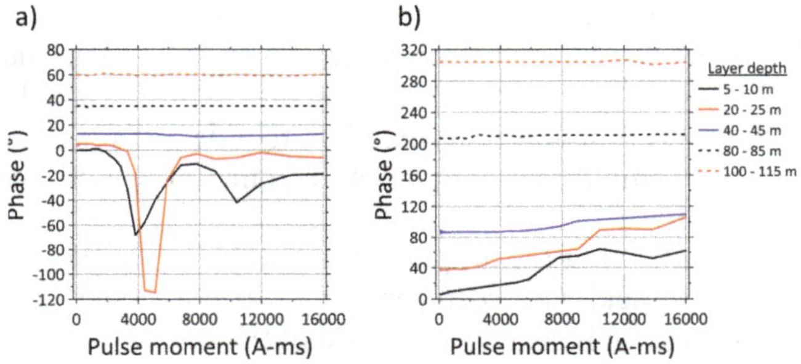


**Figure 4.9.** Phase of the MRS signal versus pulse moment computed considering different depths of a 5-m-thick water-saturated layer: (a) 100-Ω.m-half-space; (b) 10-Ω.m-half-space.

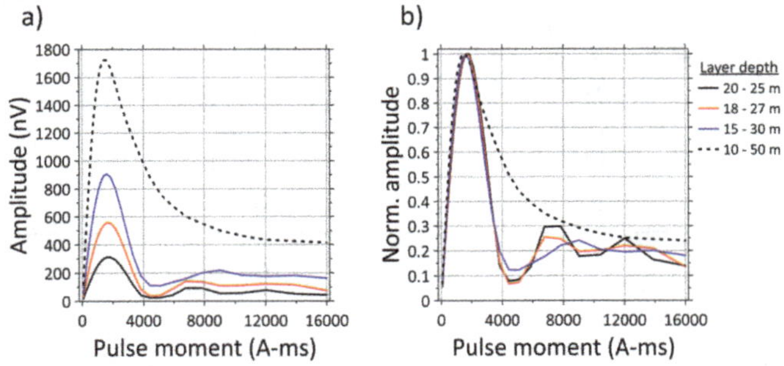


**Figure 4.10.** Example of the MRS response (amplitude versus pulse moment) corresponding to different water-saturated layers: (a) amplitude; (b) normalized amplitude.

Figure 4.9 shows that the phase shift of the MRS signal caused by the electrically conductive subsurface is a function of both the resistivity and the depth to a water-saturated layer.

Figure 4.10(a) shows the $v1(q)$ function computed assuming four models. Each model comprises one layer of different thicknesses with the water content $w = 0.4$. The depth $z$ and the thickness $\Delta z$ of these layers are selected so that the maximum of the signal corresponds to the same value of the pulse moment. The normalized MRS amplitudes presented in figure 4.10(b) show that three models have equal shapes. The shape of the fourth layer (black dashed line) is different. Note that the amplitude is directly proportional to the water content and signals with equal shapes can be scaled by adjusting the water content for rendering the signals equal. Different hydraulic models that produce equal MRS signals are termed the equivalent models. MRS measurements do not allow for distinguishing between equivalent models.

The equivalent models have different values of $w$ and $\Delta z$ and so $w$ and $\Delta z$ cannot be resolved with the MRS. However, a product of the water content and the thickness $w \times \Delta z$ is a more stable parameter characterizing hydraulic models. For example, table 4.1 shows parameters corresponding to the models generating signals presented in figure 4.10. Note that the model 10–50 m is not an equivalent model.

Figures 4.11 and 4.12 demonstrate the effect of the inclination of the Earth's magnetic field on the MRS signal. The inclination has little effect on the shape of the $v1(q)$ function corresponding to a shallow layer (figure 4.11), but the maximum amplitude is 1.7 times larger in the horizontal Earth's magnetic field than in the vertical one.

The shape of the $v1(q)$ function corresponding to a deep layer is more sensitive to the inclination, but the amplitude changes less in comparison with the shallow layer (figure 4.12).

Additional factors affecting the amplitude of the MRS signal are the magnitude of the Earth's magnetic field with corresponding Larmor frequency, and the water temperature. The amplitude is a square function of the Larmor frequency and hence, the geographical location investigated with the MRS area is an important factor to

**Table 4.1.** The product of the water content and the thickness of different models.

| Layer | $\Delta z$ $(m)$ | $w$ | $w \times \Delta z$ $(m)$ |
|---|---|---|---|
| 20–25 m | 5 | 0.4 | 2.0 |
| 18–27 m | 9 | 0.22 | 2.02 |
| 15–30 m | 15 | 0.14 | 2.06 |
| 10–50 m | 40 | 0.07 | 2.9 |

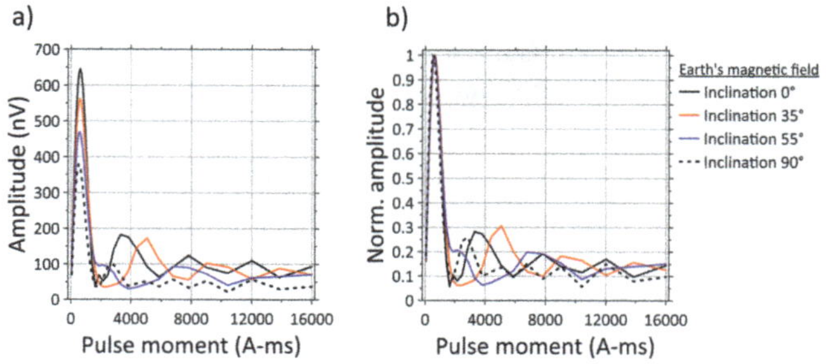


**Figure 4.11.** Amplitude of the MRS signal versus pulse moment generated by a water-saturated layer located between 5 and 10 m (5-m-thick) computed considering different inclinations of the Earth's magnetic field: (a) amplitude; (b) normalized amplitude.

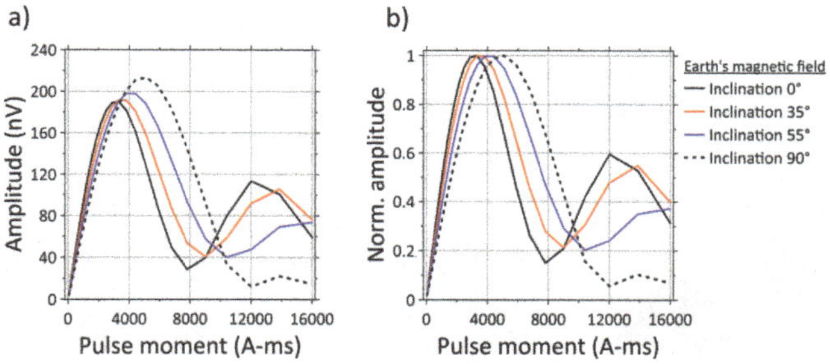


**Figure 4.12.** Amplitude of the MRS signal versus pulse moment generated by a water-saturated layer located between 40 and 45 m (5-m-thick) computed considering different inclinations of the Earth's magnetic field: (a) amplitude; (b) normalized amplitude.

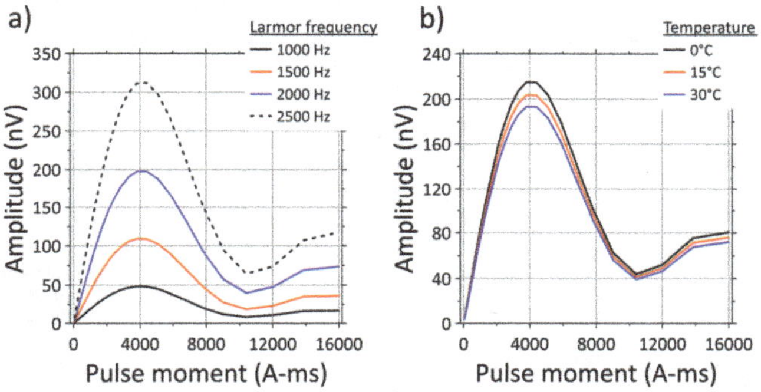


**Figure 4.13.** Additional factors affecting amplitude of the MRS signal: (a) the Larmor frequency; (b) water temperature. For demonstration, the MRS signal was computed considering a water-saturated layer between 40 and 45 m.

consider. Around the globe, the Larmor frequency varies from 800 to 2600 Hz and, for example, the amplitude of the MRS signal generated by groundwater at 1000 Hz is 6.5 times smaller than that at 2500 Hz (figure 4.13(a)). Figure 4.13(b) shows that the water temperature has little effect and the amplitude at 30°C is only 8% lower than at 0°C.

Thus, the Earth's magnetic field is an important factor affecting MRS measurements. At the scale of MRS field setup, the Earth's magnetic field may be not constant. It changes on time scales from milliseconds to millions of years. Shorter time scales arise from currents in the ionosphere and magnetosphere, and some changes can be traced to geomagnetic storms or daily variations in currents. Thus, an MRS can be started with one value of the Earth's magnetic field and finished with another one. When rocks have low magnetic permeability, the Earth's magnetic field varies only in time (figure 4.14(a)). Otherwise, the Earth's magnetic field can be changed by rocks and varies both in time and in space (figure 4.14(b)). In both cases, the MRS instrument cannot be tuned to the Larmor frequency and measurements are carried out under off-resonance conditions.

Figures 4.15 and 4.16 demonstrate variations of the MRS response computed considering near-resonance (red line) and off-resonance conditions.

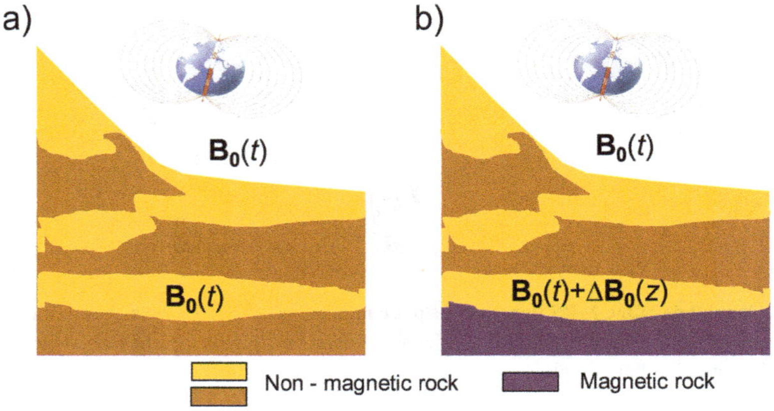


**Figure 4.14.** (a) If the subsurface is composed of non-magnetic geological materials, then the Earth's magnetic field is constant but may vary in time. (b) If the subsurface contains magnetic materials, then the Earth's magnetic field is not constant and also may vary in time.

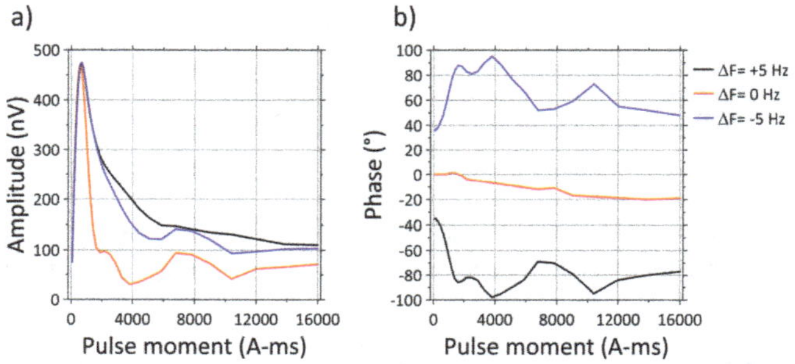


**Figure 4.15.** MRS response generated by a water-saturated layer located between 5 and 10 m (5-m-thick) computed considering different frequency shifts between the Larmor frequency and the current pulse frequency: (a) amplitude; (b) phase.

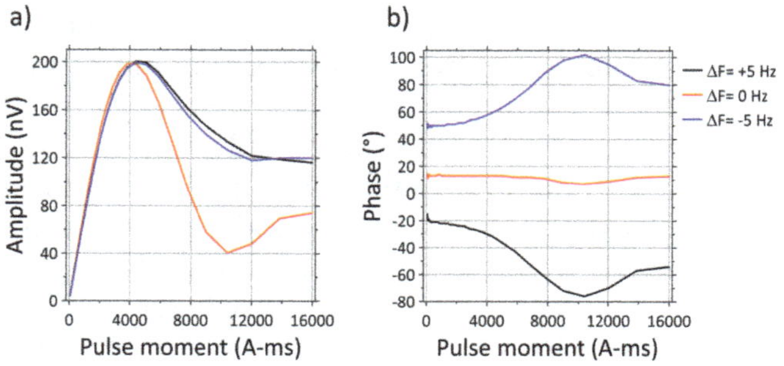


**Figure 4.16.** MRS response generated by a water-saturated layer located between 40 and 45 m (5-m-thick) computed considering different frequency shifts between the Larmor frequency and the current pulse frequency: (a) amplitude; (b) phase.

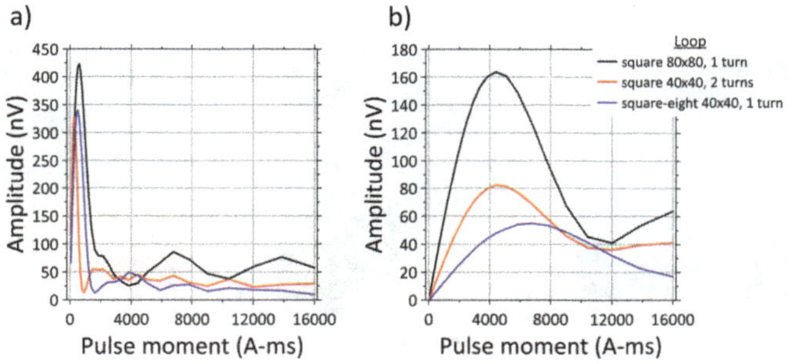


**Figure 4.17.** Amplitude of the MRS signal versus pulse moment computed considering different measuring loops: (a) a water-saturated layer located between 5 and 10 m (5-m-thick); (b) a water-saturated layer located between 40 and 45 m.

The amplitude and the shape of MRS response $v1(q)$ is also dependent on the loop surface area. Figure 4.17 shows examples of the MRS signal computed considering different loops.

## 4.3 The sensitivity of the MRS loop

MRS loops are large and even when the subsurface is homogeneous, the contribution of water to measured MRS signal depends on the pulse moment and water location under the loop. Figure 4.18 shows 1D kernels corresponding to different MRS loops. Each graph represents the amplitude of MRS signal (color scale) versus pulse moment generated by a 1-cm horizontal layer of bulk water ($w$ = 1.0) at different depths. Presented examples show the sensitivity of the MRS loop to water in the subsurface. The circular loop with the largest surface area allows larger signals and larger depth of investigation. The square and the figure-eight loops have equal surface areas, but the depth of investigation is larger with the square loop.

Figure 4.19 shows the maximum amplitude (color scale) generated by a 10 × 10 × 5 m rectangular box (cell) filled with bulk water. The set of cells make a 5-m-thick layer at the depths of 10 and 40 m. For computing, we assume the inclination of the Earth's magnetic field 55°N. Thus, the sensitivity of the MRS loop varies not only with depth but also laterally. For example, the same cell in the southern part of the investigated area produces four to five-fold larger signal than that in the northern part. Note that the contribution of different areas depends on the inclination of the Earth's magnetic field (Legchenko 2013).

Figures 4.20 and 4.21 show the sensitivity of the figure-eight loop with the main axis oriented north–south (N–S) and east–west (E–W).

In the horizontally stratified subsurface, variable sensitivity of the MRS loop does not affect the 1D interpretation of MRS measurements. However, when the subsurface is composed of 2D or 3D formations, the 1D interpretation approximates the 2D or 3D structures by equivalent 1D models (horizontal infinite layers producing equal MRS signals). For example, figure 4.22(a) shows a 2D model and its 1D approximation (figure 4.22(b)).

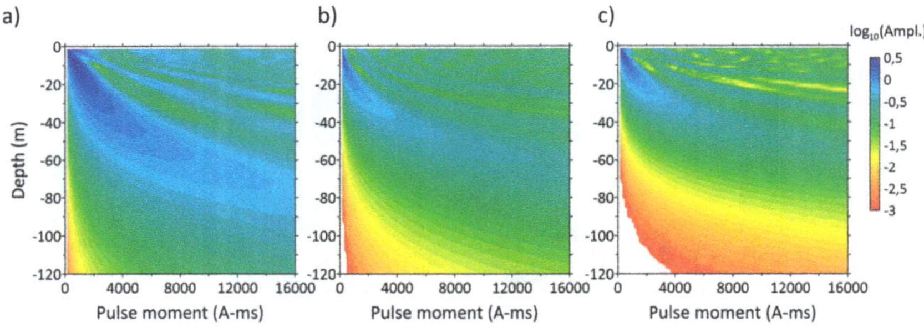


**Figure 4.18.** 1D kernels computed considering MRS loops: (a) a circular loop $D = 100$ m (1 turn); (b) a square loop of 40-m-side (2 turns); (c) a square figure-eight loop of 40-m-side and 2 turns for each square.

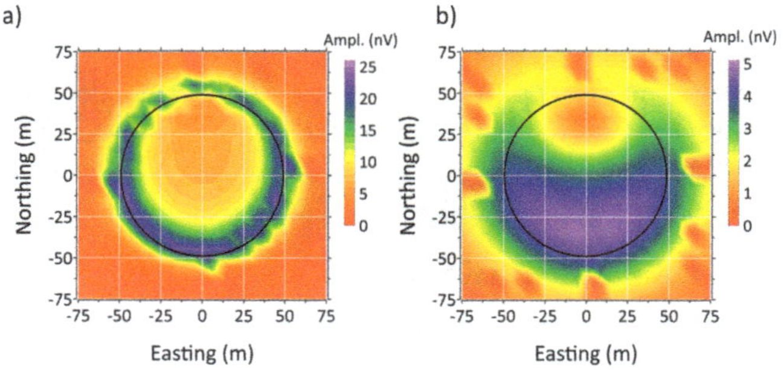


**Figure 4.19.** Amplitude of the MRS signal generated by a 10 × 10 × 5 m rectangular cell as a function of the cell position around the circular loop ($D = 100$ m, 1 turn): (a) depth of 10 m; (b) depth of 40 m.

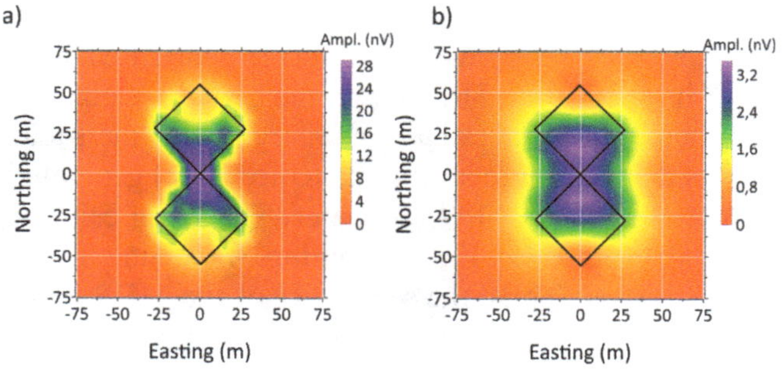


**Figure 4.20.** Amplitude of the MRS signal generated by a 10×10×5 m rectangular cell as a function of the cell position around the figure-eight loop (40-m-side, 1 turn) with the main axis oriented in the N–S direction: (a) depth of 10 m; (b) depth of 40 m.

The following examples illustrate errors caused by the 1D interpretation of signals produced by 2D models. Figure 4.23 shows a 2D model composed of one water-saturated layer ($w = 0.4$) between 40 and 45 m in the northern part of the area investigated with the MRS loop (figure 4.23(a)). The sensitivity of the circular loop ($D = 100$ m) at 40 m is shown in figure 4.23(b). The second 2D model is composed of similar water-saturated layer but in the southern part of the investigated area (not shown).

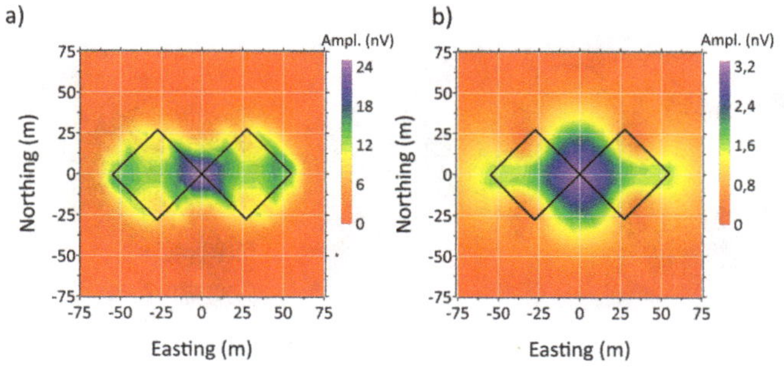


**Figure 4.21.** Amplitude of the MRS signal generated by a 10 × 10 × 5 m rectangular cell as a function of the cell position around the figure-eight loop (40-m-side, 1 turn) with the main axis oriented the E–W direction: (a) depth of 10 m; (b) depth of 40 m.

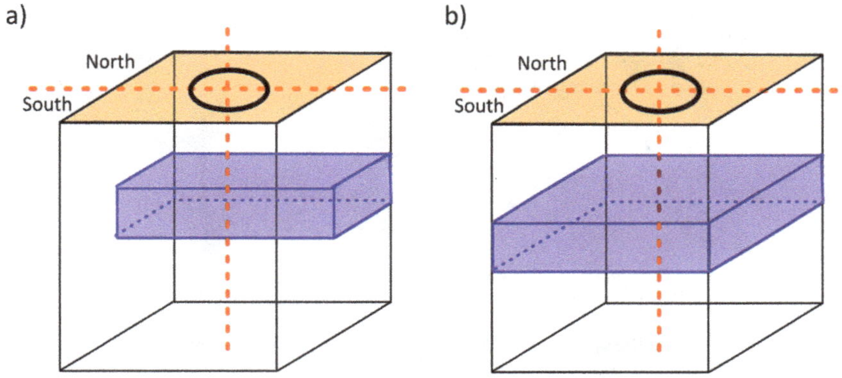


**Figure 4.22.** (a) A 2D model; (b) an equivalent 1D model.

Figure 4.24(a) shows amplitude of the MRS signal versus pulse moment computed considering these 2D models (circles) and the amplitude corresponding to the equivalent 1D models (solid lines). Figure 4.24(b) shows the initial 2D model (blue line) and the equivalent 1D models (black and red lines) corresponding to the northern and southern position of water in the 2D models.

Because of the inhomogeneous sensitivity of the MRS loop, the depth, the thickness and the water content corresponding to the initial 2D hydraulic models are shown inaccurately by the equivalent 1D models. It can be shown that the errors depend on the position of 2D and 3D water-saturated structures.

The water content and the thickness of water-saturated layers allow for estimating the equivalent water column corresponding to the volume of water in the subsurface per surface unit

$$H_w = w \times \Delta z. \tag{4.22}$$

Table 4.2 shows the water column corresponding to the initial 2D models and the equivalent 1D models (figure 4.24(b)).

In these examples, the 1D models underestimate the water column by a factor of two. However, in the initial 2D models the water-saturated formations occupy only

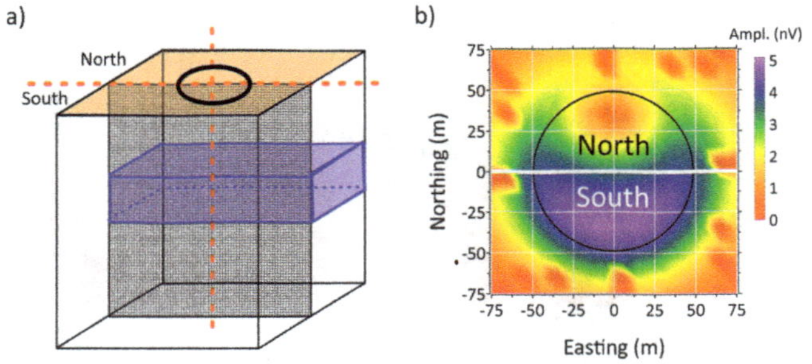


**Figure 4.23.** Example of a 2D model. (a) The water-saturated layer is located in the northern part of the subsurface relative to the axis of the measuring loop (circular, D = 100 m); (b) amplitude of the MRS signal generated by a 10 × 10 × 5 m rectangular cell located at the depth of 40 m as a function of the cell position around the measuring loop.

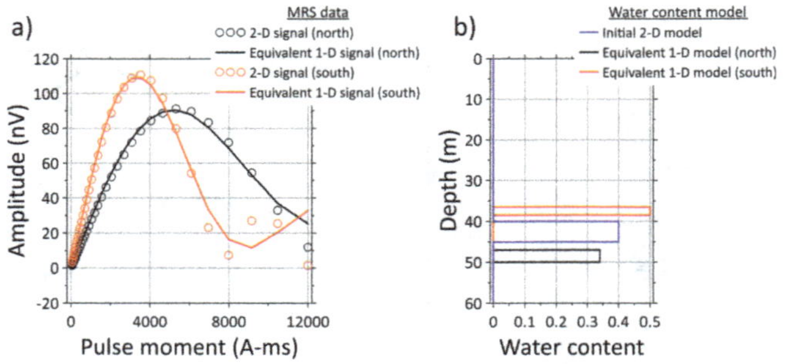


**Figure 4.24.** (a) Amplitude of the synthetic MRS signal versus pulse moment computed considering location of the 2D model (red and black circles) and signals computed the equivalent 1D models (red and black lines); (b) the initial 2D model (blue line) and the equivalent 1D models (red and black lines).

**Table 4.2.** The product of the water content and the thickness of the initial 2D models and the equivalent 1D models.

| Model | $z$(m) | $\Delta z$(m) | $w$ | $H_w$ (m) |
|---|---|---|---|---|
| 2D (north) | 40–45 | 5 | 0.4 | 2.0 |
| 2D (south) | 40–45 | 5 | 0.4 | 2.0 |
| Equivalent 1D (north) | 47–50 | 3 | 0.34 | 1.02 |
| Equivalent 1D (south) | 36.5–38.5 | 2 | 0.5 | 1.0 |

half of the investigated with MRS area. Averaged over the surface area of the loop, all four models show an equal amount of water below the loop.

## 4.4 The depth of investigation

The depth of investigation with the MRS method comprises the maximum depth of water detection and the resolution. The resolution of MRSs is linked with the inverse problem and will be investigated in the next chapter. The maximum depth of water detection is limited by the amplitude of the measurable MRS signal (S/N>1). It depends on many factors and for each configuration of measuring, the maximum depth should be estimated using forward modeling. An estimate can be done by defining the loop, the electrical resistivity of the subsurface represented by a half-space and the Earth's magnetic field. Assuming a 1-m-thick test-layer of bulk water and the 5-nV-threshold for the MRS instrument, the maximum depth of water detection is defined as the depth of the test-layer at which the maximum MRS signal is equal to 5 nV. The maximum depth is computed versus the maximum pulse moment that depends on the design of MRS instrument.

Modeling examples (figures 4.25–4.27) show that with a circular loop ($D$ = 100 m) and considering the practically available maximum pulse moment of $q$ = 15 000 A-ms, the test-layer can be detected at a depth of 160 m. However, the maximum depth of

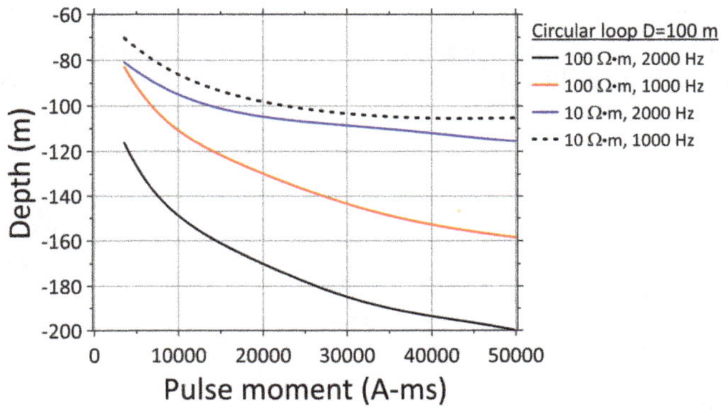


**Figure 4.25.** The depth of a 1-m-thick test-layer ($w$ = 1), which generates the maximum MRS signal of 5 nV versus the maximum pulse moment. Modeling was performed considering a circular loop ($D$ = 100 m), the half-space resistivity of 10 and 100 Ω m, inclination of the Earth's magnetic field of 55° and the Larmor frequency of 1000 and 2000 Hz.

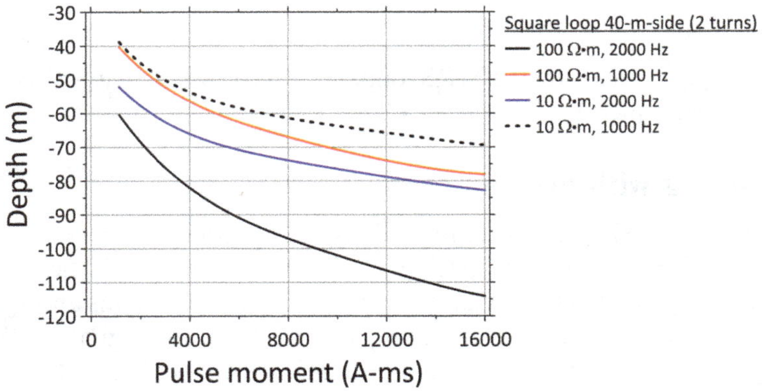


**Figure 4.26.** The depth of a 1-m-thick test-layer ($w$ = 1), which generates the maximum MRS signal of 5 nV versus the maximum pulse moment. Modeling was performed considering a square loop of 40-m-side with two turns, the half-space resistivity of 10 and 100 Ω m, inclination of the Earth's magnetic field of 55° and the Larmor frequency of 1000 and 2000 Hz.

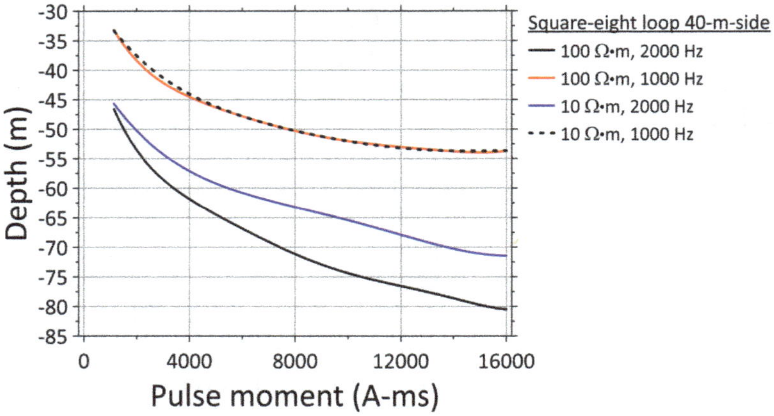


**Figure 4.27.** The depth of a 1-m-thick test-layer ($w = 1$), which generates the maximum MRS signal of 5 nV versus the maximum pulse moment. Modeling was performed considering a square figure-eight loop of 40-m-side, the half-space resistivity of 10 and 100 Ω m, inclination of the Earth's magnetic field of 55° and the Larmor frequency of 1000 and 2000 Hz.

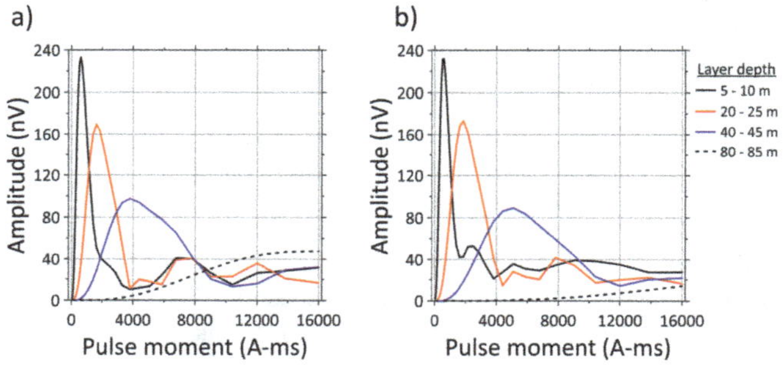


**Figure 4.28.** Amplitude of the SE signal versus pulse moment computed considering different depths of a 5-m-thick water-saturated layer: (a) 100-Ω.m-half-space; (b) 10-Ω.m-half-space.

detection strongly depends on the MRS loop, the electrical resistivity of the subsurface and the Larmor frequency.

## 4.5 Measuring with the spin echo

Figure 4.28 shows SE signals computed using equation (4.20) and assuming a circular loop ($D$ = 100 m), inclination of the Earth's magnetic field of 55°, the Larmor frequency of 2000 Hz, delay between the pulses $\tau_d$ = 90 ms and $T_2$ = 400 ms. The diffusion effect was neglected. Comparison of the SE signals with the FID signals (figure 4.7) shows that the shape of SE signals is similar to that of FID signals, but the amplitude is about twice as small.

Figure 4.29 shows SE and FID kernels.

The max depth of water detection (figure 4.30) is smaller with SE measurements. As the SE measuring procedure requires a twice as short pulse, the maximum pulse

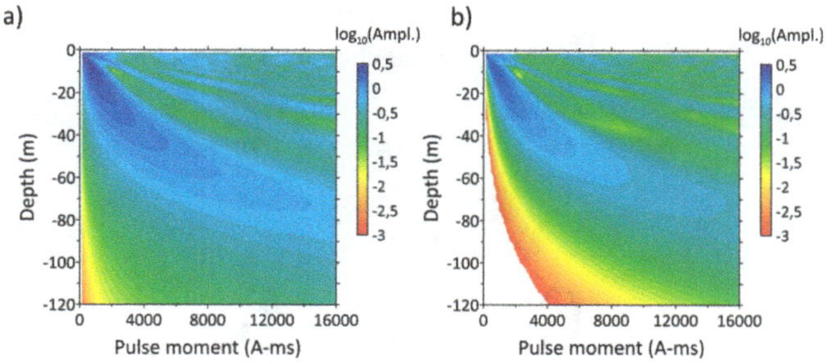


**Figure 4.29.** 1D kernels computed considering the circular loop $D = 100$ m (1 turn), a 100-Ω.m-half-space and different measuring procedures: (a) FID; (b) SE.

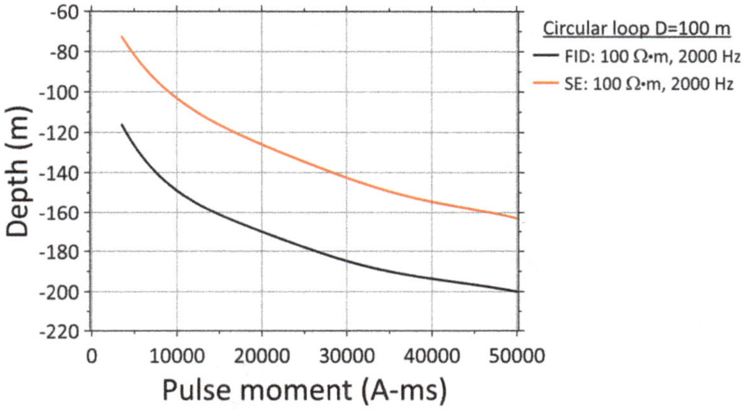


**Figure 4.30.** The depth of a 1-m-thick test-layer ($w = 1$), which generates the maximum MRS signal of 5 nV versus the maximum pulse moment. Modeling was performed considering the FID and SE measuring procedures.

moment is twice as small when using SE. Thus, assuming available $q_{max} = 15000$ A-ms for FID measurements and 7500 A-ms for SE measurements, the maximum depth of the test-layer detection is estimated as 160 m (FID) against 90 m (SE).

## Bibliography

Legchenko A 2013 *Magnetic Resonance Imaging for Groundwater* (London: Wiley-ISTE)

# Chapter 5

# MRS inversion

Figure 5.1 presents a flowchart of the magnetic resonance sounding (MRS) inverse modeling. The linear filter is computed following the same procedure as for the forward modeling. A MRS data set contains experimental data. Note that for investigating properties of MRS inversion, a synthetic data set can be used instead of the experimental data. The solution of the MRS inverse problem provides the water content and the relaxation times in the subsurface.

The goal of the MRS inversion is to build an inverse model of the water content and of the relaxation times that minimizes the discrepancy between the experimental data and the theoretical signal computed considering the inverse model. An example of the algorithm of the iterative optimization is shown in figure 5.2. At each iteration, the inverse model is revised so that the discrepancy becomes smaller than the experimental errors.

The discrepancy between experimental and theoretical signals can be quantified using the root-mean-square error

$$\text{RMSE} = \sqrt{\frac{1}{N}\sum_{i=1}^{N}(v_i - \tilde{v}_i)^2} \tag{5.1}$$

where $\tilde{v}_i$ and $v_i$ are respectively measured and theoretical amplitudes.

The MRS inverse problem is done by solving the algebraic equation

$$\mathbf{Ax} = \mathbf{b} \tag{5.2}$$

where $\mathbf{A} = [a_{i,j}]$ is a rectangular matrix representing the kernel an integral equation, $i = 1,2,.., i, .., I$ is the number of pulse moments and $j = 1,2,.., j, .., J$ is the number of elements in the inverse model, $\mathbf{x} = [x_j]$ is the model and the vector $\mathbf{b} = [b_i]$ contains measured signals. When using the FID measurements the matrix $\mathbf{A}$ approximates equation (2.19). Measurements of the SE signal requires approximating equation (2.23).

doi:10.1088/978-0-7503-3155-5ch5  

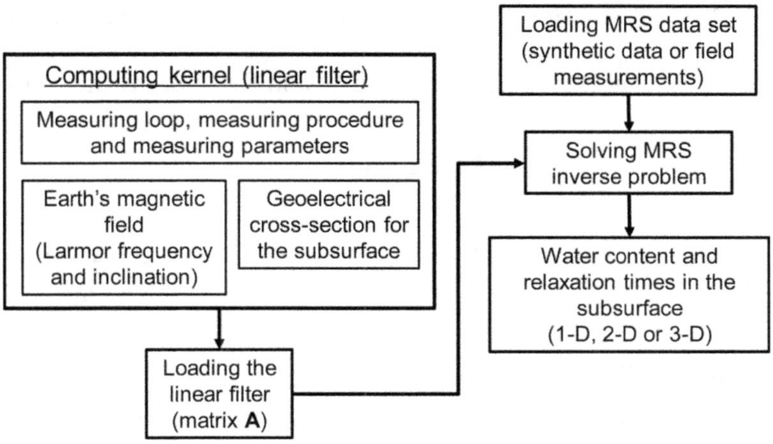


**Figure 5.1.** A flowchart of the MRS inverse modeling procedure.

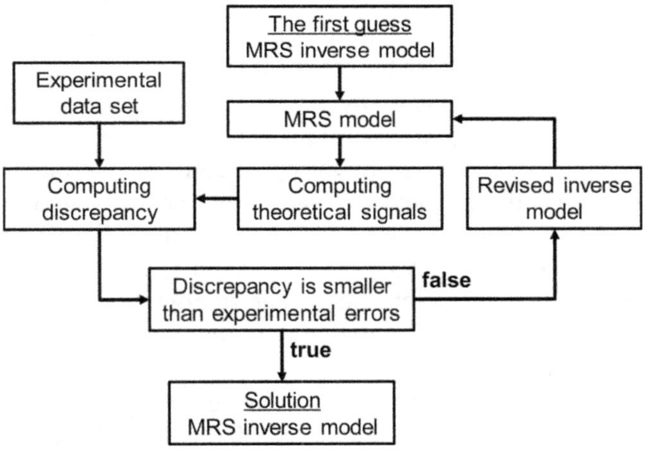


**Figure 5.2.** An algorithm for solving the MRS inverse problem.

In equation (5.2), inversion of the FID1 measurements for water content (equation 4.15) assumes the substitution

$$\begin{cases} \mathbf{x} = \mathbf{Tw} \\ \mathbf{b} = \mathbf{v1} \end{cases} \tag{5.3}$$

where the matrix $\mathbf{T}$ is computed with the corresponding relaxation factors $RF_j$ (equation 4.17).

Equation (4.18) corresponds to the FID2 inversion for $T_1$. In equation (5.2), the solution vector $\mathbf{x}$ and the data set $\mathbf{b}$ becomes

$$\begin{cases} \mathbf{x} = \mathbf{Tw} \\ \mathbf{b} = \mathbf{v2} \end{cases} \tag{5.4}$$

where the matrix **T** is computed with the corresponding relaxation factors $RF_j$ (equation 4.19).

Equation (4.20) corresponds to the SE inversion for water content. In equation (5.2), the solution vector **x** and the data set **b** are

$$\begin{cases} \mathbf{x} = \mathbf{Tw} \\ \mathbf{b} = \mathbf{se\tau} \end{cases} \tag{5.5}$$

where the matrix **T** is computed with the corresponding relaxation factors $RF_j$ (equation 4.21).

## 5.1 The equivalence problem

The solution of the MRS inverse problem comprises representing the subsurface by the discrete elements (figure 4.2). In the 1D case, each element is a horizontal infinite layer. In the 2D case, it is a rectangular box infinite in one direction. In the 3D case, the subsurface is represented by a set of finite homogeneous boxes. Note that the solution can be built using one, two or many elements of equal or different shapes and sizes. Assumptions on the solution shape depend on the complexity of the subsurface and prior information about the subsurface.

The MRS inverse problem has many solutions that fit experimental data within experimental error. All these solutions are the equivalent solutions. A number of the equivalent solutions can be reduced by applying constrains. Constraints should be justified by knowledge of the physical nature of the measured parameters, the geological structure of the subsurface and other considerations. For example, it is known that the water content can vary between 0 and 1, and the equivalent solutions containing $w < 0$ or $w > 1$ can be excluded. The smoothness constraint is also often applied in geophysics. Based on the variety of criteria, one equivalent solution can be selected as the best solution.

Let us investigate the equivalence problem using a simple 1D model. The model comprises a one 10-m-thick layer at a depth of 20 m with the water content $w = 0.1$ (figure 5.3(a), black dashed lines). We use a coincident (Tx/Rx) square loop with the side length of 80 m and compute the synthetic MRS signal corresponding to this model. Then, we contaminate the signal by a normally distributed random noise with the standard deviation $\sigma = 8.5$ nV (figure 5.3(b), black crosses). Inversion of this data set applying the constraint of a non-negative water content provides many equivalent models, and we show five of them. The theoretical signal corresponding to each equivalent model fits the synthetic data with the root-mean-square error RMSE $\approx 6.1$ nV. The inverse models 1 and 2 contain only one layer each, the inverse model 3 comprises two layers, and the inverse models 4 and 5 are multi-layer models with different numbers of layers. Figure 5.3(b), shows the theoretical signals corresponding to each of these inverse models (red lines).

Figure 5.4(a) shows inverse models (solid lines) provided by inversion of five data sets. The data sets correspond to the synthetic signal computed after the initial model (black dashed lines) contaminated by different realizations of a normally distributed random noise with the standard deviation $\sigma = 8.5$ nV. For each data set, one multi-layer model

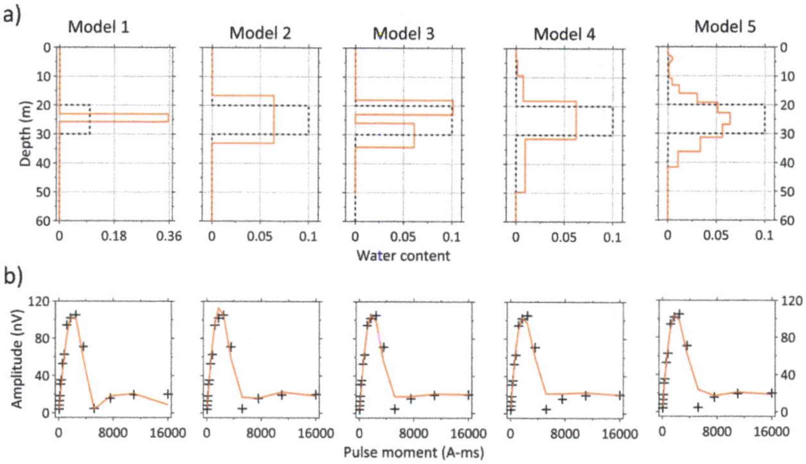


**Figure 5.3.** (a) Examples of the equivalent models (red lines) and the initial model (black dashed lines). (b) The MRS signal (crosses) and the theoretical signal corresponding to each inverse model (red lines). All the theoretical signals fit the MRS signal with the same error RMSE = 6.1 nV.

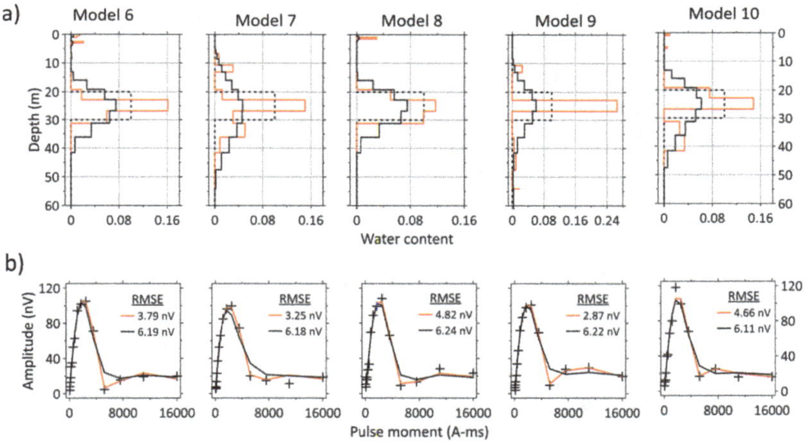


**Figure 5.4.** (a) Equivalent models: smooth solution (black solid lines) and the best fit solution (red solid lines). The initial model is shown by the black dashed lines. (b) The MRS signals correspond to each model (crosses) and the theoretical signals correspond to the inverse models (black and red solid lines).

provides the best fit of the data (red solid lines) and one inverse model is a smooth model (black solid lines). The smooth models fit data with approximately equal misfits. Figure 5.4(b) shows the synthetic data sets (crosses) and the theoretical signals corresponding to these inverse models (solid lines).

Figures 5.3 and 5.4 show that because of the equivalence problem, neither water content nor thickness of a water-bearing formation can be accurately derived from MRS inversion. Assumptions on the solution shape allow for getting stable solution, but the assumptions have to be consistent with the subsurface structure. However, if one parameter, the water content or the thickness, would be known, then the second parameter can be estimated with a much better accuracy. Also, MRS is an efficient tool for following changes in the water saturation by performing inversion of all soundings using the same constraints.

In a 3D case, MRS allows for estimating the volume of water in the subsurface as

$$V = \sum_{j=1}^{J_H} (w_j \times V_j) \tag{5.6}$$

where $w_j$ is the water content in the cell $j$, $V_j = \Delta x_j \times \Delta y_j \times \Delta z_j$ is the cell volume and $J_H$ is the number of cells considered for computing the water column. The water volume is the most stable parameter estimated with MRS. In the 1D case, the water volume is estimated by computing an equivalent water column $H$ by assuming the surface area $\Delta S = 1\ \mathrm{m}^2$

$$H = \sum_{j=1}^{J_H} (w_j \times \Delta z_j). \tag{5.7}$$

Let us verify the stability of the water column estimate by computing $H$ considering the models shown in figures 5.3 and 5.4. Table 5.1 shows $H$ computed using equation (5.7).

Figure 5.5 shows the $H$ estimate presented in table 5.1. The red line shows the true value corresponding to the initial model and the black solid line shows the mean value computed using inverse models 2–16. For this data set, the mean value suggests that the inversion overestimate $H$ with a systematic error (10%). This error

**Table 5.1.** The water column computed considering each inverse model (crosses) shown in the previous figures.

| *N* | *Model* | *MRS water column* (m) |
|---|---|---|
| 1 | Initial model | 1.0 |
| 2 | Model 1 | 0.962 |
| 3 | Model 2 | 1.059 |
| 4 | Model 3 | 1.008 |
| 5 | Model 4 | 1.067 |
| 6 | Model 5 | 1.069 |
| 7 | Model 6 the best fit | 0.998 |
| 8 | Model 6 smooth | 1.100 |
| 9 | Model 7 the best fit | 1.167 |
| 10 | Model 7 smooth | 1.115 |
| 11 | Model 8 the best fit | 1.113 |
| 12 | Model 8 smooth | 1.085 |
| 13 | Model 9 the best fit | 1.344 |
| 14 | Model 9 smooth | 1.116 |
| 15 | Model 10 the best fit | 1.186 |
| 16 | Model 10 smooth | 1.128 |

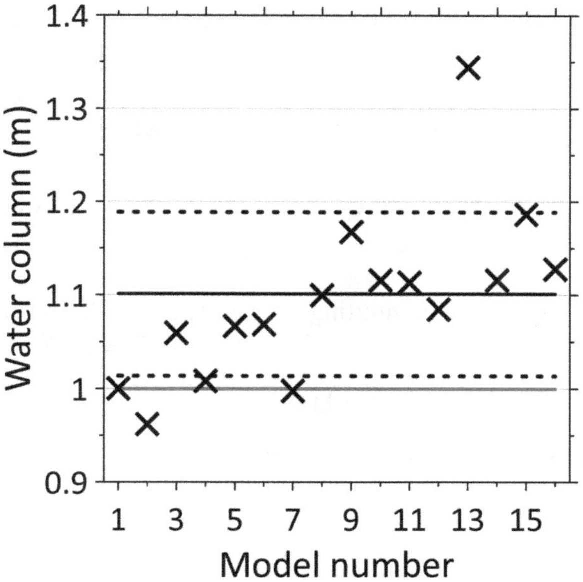


**Figure 5.5.** The water column computed considering each inverse model (crosses) shown in the previous figures. The red line shows the true water column (1.0 m), the black solid line shows the mean of the water column (1.1 m), and the dashed lines show the standard deviation of the water column estimate (+/− 0.0875 m).

is due to the filtering of the signal, which causes the loss of the signal amplitude. Making the band-pass of the filter wider may help to solve this problem, but the noise influence will also be larger. The standard deviation around the mean (0.0875 m) shows dispersion of the $H$ estimate around the mean. Thus, the water column is estimated as $H = 1.1 \pm 0.0875$ m, which is more accurate than the estimates of $w$ and $\Delta z$ shown in figures 5.3 and 5.4.

## 5.2 The resolution

MRS inversion comprises solving equation (5.2)

$$\mathbf{x} = \mathbf{A}^{-1}\mathbf{b} \tag{5.8}$$

where $\mathbf{A}^{-1}$ is the inverse of the matrix $\mathbf{A}$, $\mathbf{x}$ is the water content model and $\mathbf{b}$ is an experimental data set. Both stability and accuracy of the inversion depend on the properties of the matrix $\mathbf{A}$, which is a product of discretization of the integral equation corresponding to the method of measuring.

Inversion of FID and SE measurements shows a lot of similarity, and we focus the attention on the inversion of the FID1 measurements for water content. A 1D discretization of equation (4.1) requires the defining of the number of layers $J$ in the inverse model, the depth $z_j$ and the thickness $\Delta z_j$ of each layer, the number ($I$) and values of the pulse moment ($q_i$) used for measuring MRS signals. Selection of the parameters of discretization can be done using different approaches, and we present one of them often used in practice.

We cut the subsurface into $J$ layers (figure 4.2) so that their thickness increases downwards

$$\Delta z_j = z_{j+1} - z_j, \quad z_{\max} = \sum_{j=1}^{J} \Delta z_j \tag{5.9}$$

where $\Delta z_1 \leqslant \Delta z_2 \leqslant .. \leqslant \Delta z_j \leqslant .. \leqslant \Delta z_J$ and $z_{\max}$ is the maximum depth of water detection with MRS. In the discrete inverse problem, these model layers represent the subsurface.

Selection of $\Delta z_j$ comprises computing the Pearson correlation coefficient between the columns of the matrix **A**. The correlation matrix $\mathbf{R} = [r_{i,j}]$ is

$$\mathbf{R} = \mathbf{D}\mathbf{A}^{\mathrm{T}}\mathbf{A}\mathbf{D} \tag{5.10}$$

where **D** is a diagonal matrix with the elements

$$d_{j,j} = \frac{1}{\sqrt{\sum_{i=1}^{I} a_{i,j}^2}}. \tag{5.11}$$

Then, $\Delta z_j$ are selected so that the correlation coefficient between the neighboring layers is equal to a fixed value $r$

$$r = r_{j,j+1} = r_{j+1,j+2} = \ldots = r_{J-1,J}. \tag{5.12}$$

Modeling results show that for keeping the correlation coefficient between the layers constant, the thickness of these layers must be increased with depth. Figure 5.6 shows thickness of model layers versus depth computed assuming a square 100-m-side loop, the maximum pulse moment of 16 000 A-ms and different values of $r$. This example shows that the number of layers $J$ is linked with the correlation coefficient $r$. Thus, selection of $J$ allows defining $\Delta z_j$. For each graph, the dashed line shows a fit computed as $\log_e(\Delta z_j) = az_j + b$.

Examples of MRS signal presented in chapter 4 demonstrate that the signal $v(q)$ has a maximum of the amplitude. The value of the pulse moment corresponding to this maximum can be used for showing changes in the MRS response. Figure 5.7 shows that values of the pulse moment depend on the electrical resistivity of the subsurface (half-space). Indeed, the subsurface is an electromagnetic screen that limits penetration of the electromagnetic field and causes the maximum of the signal to be obtained at a smaller depth. A third degree polynomial fit of these data approximates well computed pulse moments. Distinct changes in the signal's shape $v(q)$ affect the $\Delta z_j(z_j)$ function and suggest that the discretization should be done for each sounding considering measuring conditions.

The number of pulses $I$ is a trade-off between the fieldwork duration and the inversion accuracy. Many pulses allow a more detailed description of the MRS signal, thus improving inversion accuracy, but it increases time for measuring. An insufficient number of pulses may cause a loss of information, thus degrading

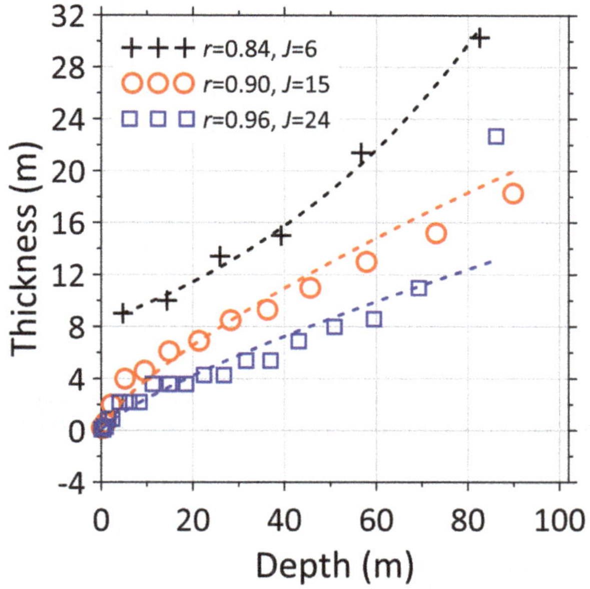


**Figure 5.6.** Thickness of the model layers ($\Delta z_j$) as a function of the correlation coefficient between the neighboring layers. The dashed lines show an exponential fit of the computed ($\Delta z_j$).

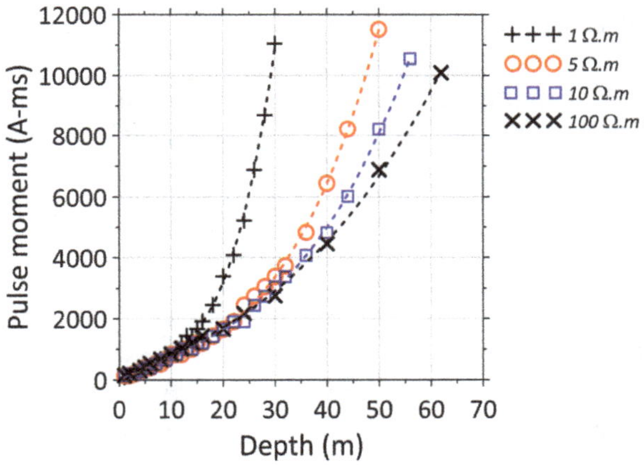


**Figure 5.7.** Values of the pulse moment corresponding to the maximum of the signal versus electrical resistivity of the subsurface. The dashed lines show a polynomial fit of the computed pulse moments.

inversion results. However, the maximum pulse moment $q_{max}$ and the number of pulses $I$ depend on the instrument parameters and time available for the fieldwork. Interpretation of field measurements is usually done when the fieldwork is over. It does not allow for varying pulse moments for the inversion of real data. In a NUMIS MRS system, the pulse moments are distributed between $q_{min}$ and $q_{max}$ after the polynomial function of the third degree, that approximately correspond to the maximum of the MRS response when $I = J$ (figure 5.8(a)).

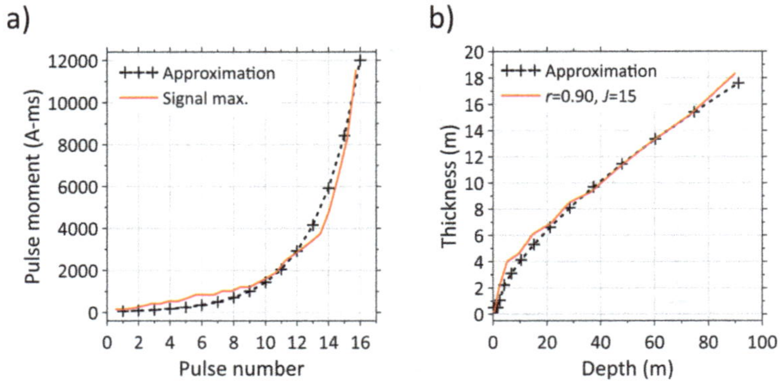


**Figure 5.8.** (a) Values of the pulse moment as a function of the pulse number computed using the polynomial approximation and assuming $J = 16$, $q_{\min} = 60$ A-ms and $q_{\max} = 12{,}000$ A-ms. (b) Thickness of the model layer versus the layer depth computed using an approximation and considering $z_{\max} = 100$ m and $J = 16$.

It is more convenient to make discretization using an approximation instead of a straightforward application of the discretization procedure with fixed correlation coefficient between the neighboring layers. For that, the thickness of model layers between 0 and $z_{\max}$ follows a distribution computed as $\log_e(\Delta z_j) = c \log_e(z_j) + d$. Figure 5.8(b) shows the thickness of model layers versus the layer depth computed using this approximation and considering $z_{\max} = 100$ m and $J = 16$.

The solution of equation (5.8) is sensitive to the parameters of the matrix **A**, which is a product of discretization of an integral equation. The matrix **A** can be computed using the procedure discussed above that comprises an estimate of $z_{\max}$ and using a different number of the model layers $J$. The singular value decomposition (SVD) allows for analyzing the solution of equation (5.8) with the given matrix **A**. For that, we present matrix **A** as a product of three orthogonal matrices **U**, **V**, and **S** (Aster *et al* 2005)

$$\mathbf{A} = \mathbf{U}\mathbf{S}\mathbf{V}^{\mathrm{T}} \tag{5.13}$$

where **U** is an $I \times I$ matrix representing the data space, **V** is an $J \times J$ matrix representing the model space and **S** is an $I \times J$ diagonal matrix with non-negative diagonal elements (singular values). The model resolution matrix $\mathbf{R}_m$ describing how well the recovered model is able to represent the original model is

$$\mathbf{R}_m = \mathbf{V}\mathbf{V}^{\mathrm{T}}. \tag{5.14}$$

If

$$\mathbf{R}_m = \mathbf{I} \tag{5.15}$$

then the model **x** in equation (5.8) will be recovered exactly and the resolution is perfect; where **I** is the identity matrix with ones on the main diagonal and zeros elsewhere.

For estimating the maximal resolution of the MRS inversion under noiseless conditions, we set the number of model layers as large as possible ($J = J_{\max}$) but keeping $\mathbf{R}_m = \mathbf{I}$. Figure 5.9(a) shows the resolution matrix $\mathbf{R}_m$ computed considering discretization with

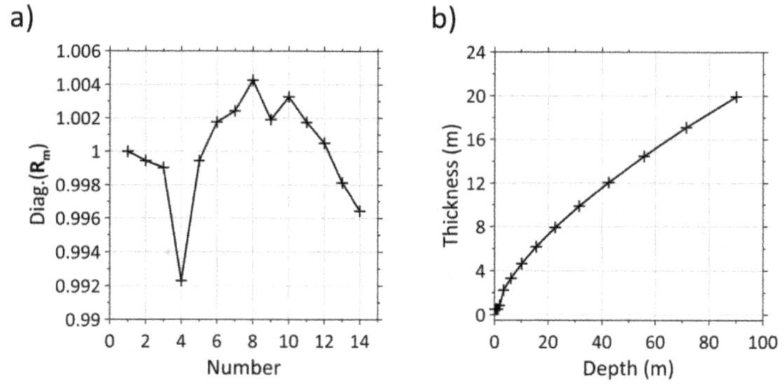


**Figure 5.9.** Example of MRS resolution computed considering discretization with $q_{max}$ = 12 000 A-ms, $J = J_{max}$ = 14: (a) the diagonal elements of the model resolution matrix; (b) the model layer thickness versus the layer depth.

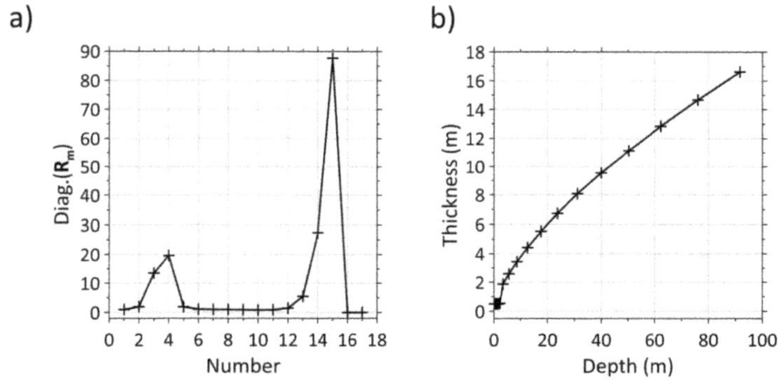


**Figure 5.10.** Example of MRS resolution computed considering discretization with $q_{max}$ = 12 000 A-ms, $J$ = 17: (a) the diagonal elements of the model resolution matrix; (b) the model layer thickness versus the layer depth.

$J$ = 14 and figure 5.10(a) with $J$ = 17. The thickness of the model layers corresponding to these two cases is shown in figures 5.9(b) and 5.10(b). In these examples, $J_{max}$ = 14 and discretization carried out with $J$ = 17 does not satisfy equation (5.15).

The major factors affecting resolution of the MRS inversion are the maximum pulse moment used for measuring, the size of the measuring loop and the electrical resistivity of the subsurface. The following numerical examples show the thickness of the model layer versus the layer depth computed assuming MRS loops made of a 400-m-long wire: a 100-m-side square loop, a 50-m-side square loop and a 50-m-side figure-eight loop (figure 5.11). These loops correspond to the typical MRS setup. For modeling, we assume the Larmor frequency of 2000 Hz, the inclination of the Earth's magnetic field of 55°N and 16 pulse moments ($I$ = 16).

These examples show that MRS can resolve a few-m-thick layers, but the resolution depends on measuring conditions. At shallow depth (0–5 m), MRS resolution is high ($\Delta z < 1$m) and degraded in larger depths. Note that at the scale of the real MRS loop, geological formations are rarely composed of thin horizontal layers and this resolution capacity is hardly achievable in practice.

The SVD can be used for computing the generalized inverse of **A**, called the Moore–Penrose pseudoinverse. The pseudoinverse solution is

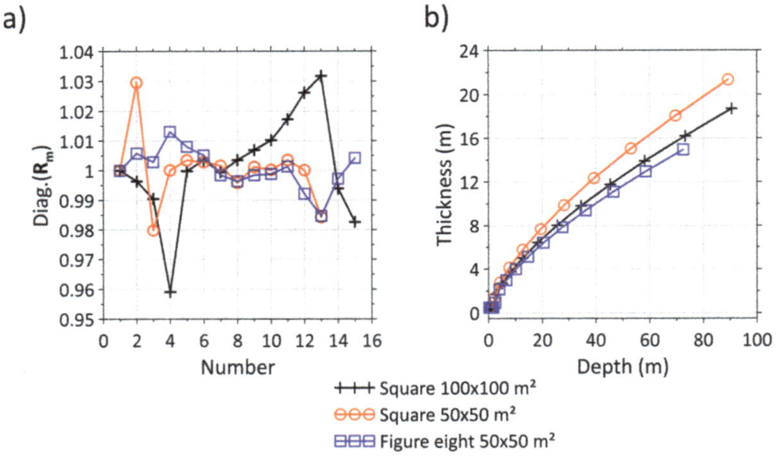


**Figure 5.11.** MRS resolution computed considering different MRS loops, a 100-Ω.m-half-space and $q_{max}$ = 12,000 A-ms: (a) the diagonal elements of the model resolution matrix; (b) the model layer thickness versus the layer depth.

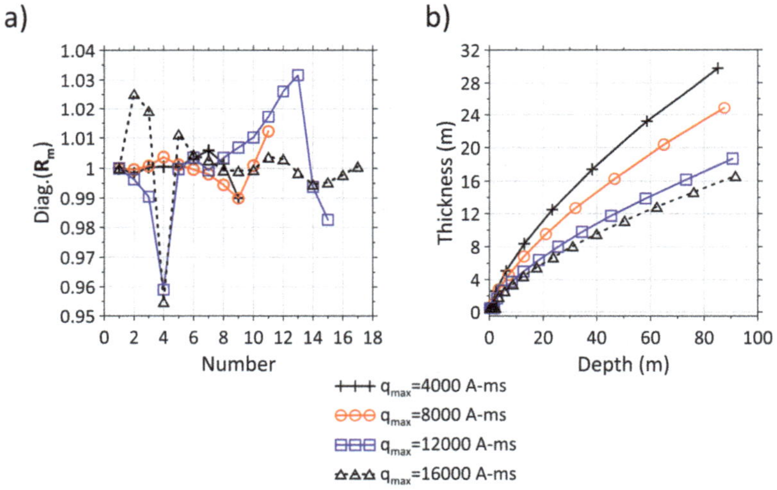


**Figure 5.12.** MRS resolution computed considering a 100-m-side square loop, a 100-Ω.m-half-space and different values of the maximum pulse moment ($q_{max}$): (a) the diagonal elements of the model resolution matrix; (b) the model layer thickness versus the layer depth.

$$\mathbf{x} = \mathbf{V}\mathbf{S}^{-1}\mathbf{U}^{\mathrm{T}}\mathbf{b}. \tag{5.16}$$

For demonstrating that if $\mathbf{R}_m = \mathbf{I}$, then the model $\mathbf{x}$ can be accurately recovered, we use a spike model which is a vector with all zero elements except one. We assume a 100-m-side square loop, a 100-Ω.m-half-space and the maximum pulse moment equal to 16 000 A-ms. The diagonal elements of the model resolution matrix with $J = 17$ is close to one, as shown in figure 5.12 by black triangles. Figure 5.14(a)

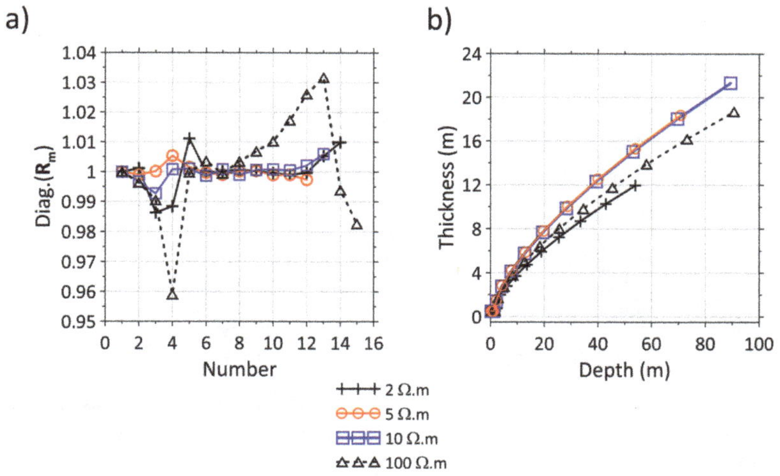


**Figure 5.13.** MRS resolution computed considering a 100-m-side square loop, $q_{max}$ = 12000 A-ms and different electrical resistivity of the subsurface (half-space): (a) the diagonal elements of the model resolution matrix; (b) the model layer thickness versus the layer depth.

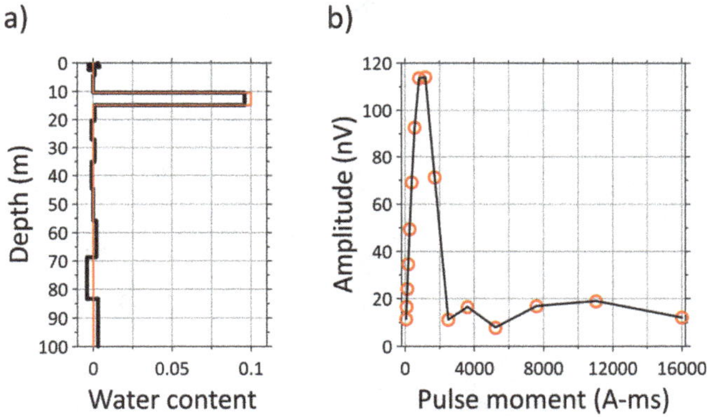


**Figure 5.14.** (a) The spike model with a shallow layer corresponding to the matrix **A** discretization (red line) and the recovered model (black line). (b) The synthetic signal (red circles) and the inversion fit with RMSE = 0.07 nV.

shows the spike model with one layer between 10.42 and 14.84 m with the water content $w_9 = 0.1$ and the model recovered by inversion. Note that the depth and the thickness of this layer coincide with one model layer defined by discretization. Figure 5.14(b) shows the MRS amplitude versus depth and the theoretical signal computed after the inverse model. A similar example is shown considering a deeper layer between 55.85 and 68.70 m with the water content $w_{15} = 0.1$. Note that the model resolution matrix predicted equal resolution for both models but we observe better accuracy of recovering the model with one shallow layer (figure 5.15).

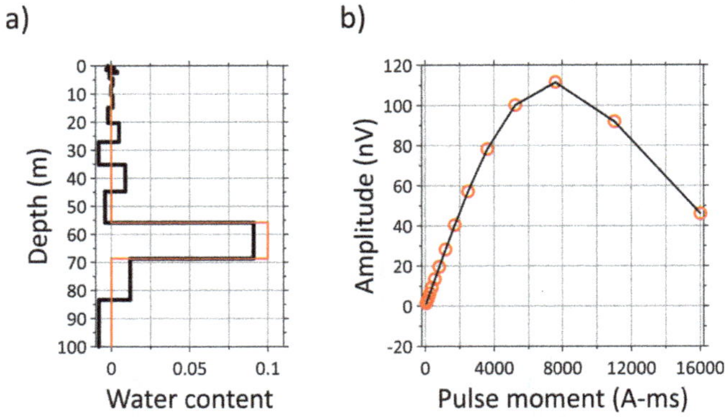


**Figure 5.15.** (a) The spike model with a deep layer corresponding to the matrix **A** discretization (red line) and the recovered model (black line). (b) The synthetic signal (red circles) and the inversion fit with RMSE = 0.08 nV.

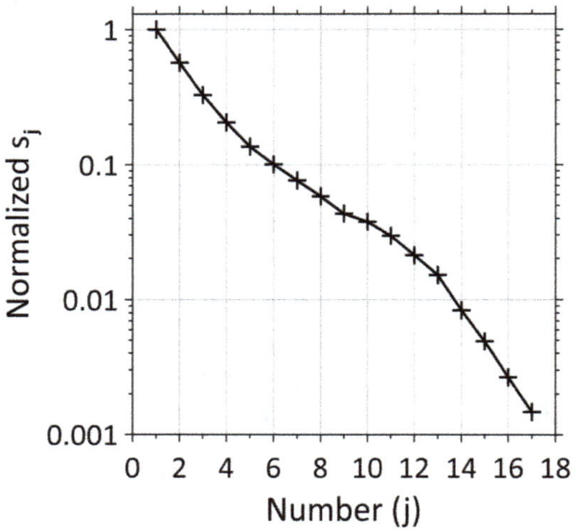


**Figure 5.16.** Normalized singular values corresponding to the discretization with $J = 17$.

The lower accuracy of recovering deep layers is caused by degraded inversion resolution indicated by smaller singular values ($s_j$) corresponding to deep model layers in the matrix **A** as shown in figure 5.16.

The condition number defined as the ratio of the largest ($s_1$) to the smallest ($s_J$) singular values of the matrix **A** allows measuring the inversion sensitivity to experimental and computational errors

$$\text{cond}(\mathbf{A}) = s_1/s_J. \tag{5.17}$$

The matrix **A** with a large condition number is more sensitive to the errors than the matrix with a small one. Figure 5.17 shows cond(**A**) versus number of the model layers $J$. The matrix **A** discretized with a smaller number of model layers has smaller condition number which renders inversion more stable.

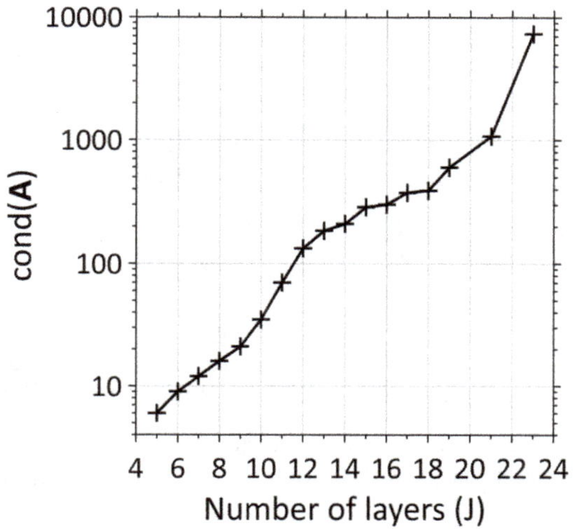


**Figure 5.17.** The condition number of the matrix **A** versus the number of model layers.

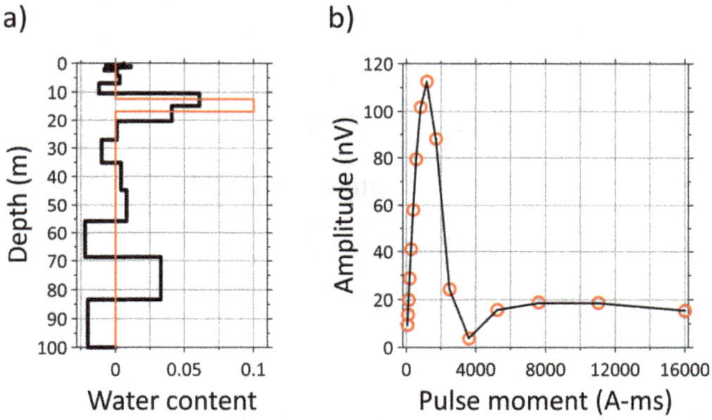


**Figure 5.18.** (a) The spike model with a shallow layer not corresponding to the matrix **A** discretization (red line) and the recovered model (black line). (b) The synthetic signal (red circles) and the inversion fit with RMSE = 0.23 nV.

At the scale of MRS loop, real geological formations may not correspond to a horizontally stratified media and the boundaries of water-bearing formations may not coincide with the boundaries of the model layers in the matrix **A**. Figures 5.18 and 5.19 show inversion of the spike models using the same matrix **A** and noiseless data sets. In these examples, the boundaries do not coincide with the layers given by the discretization. The synthetic MRS signals (figures 5.18(b) and 5.19(b), red circles) are well fitted by the theoretical signal (figures 5.18(b) and 5.19(b), black lines) computed after the inverse models (figures 5.18(a) and 5.19(a), black lines), but the models (figures 5.18(a) and 5.19(a), red lines) were recovered not as well as in the previous examples. This points out that inversion is sensitive to the discrepancy between the boundaries of the model and the model layers given by the discretization.

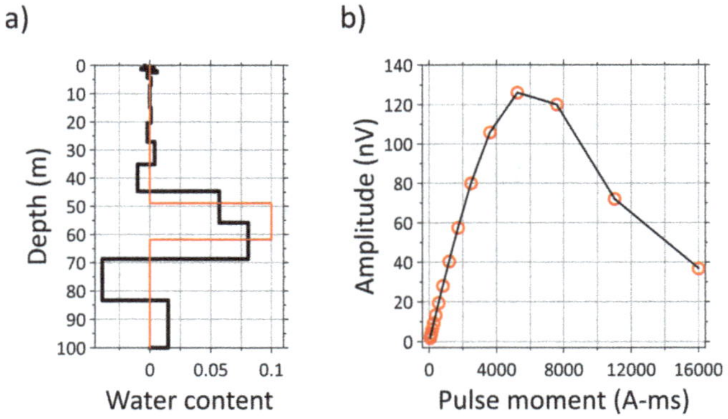


**Figure 5.19.** (a) The spike model with a deep layer not corresponding to the matrix **A** discretization (red line) and the recovered model (black line). (b) The synthetic signal (red circles) and the inversion fit with RMSE = 0.1 nV.

MRS measurements are always contaminated by experimental noise and the SVD allows for estimating inversion uncertainty for erroneous data. For that, we assume noise characterized by independent and identically distributed (i.i.d.) normal data errors $\sigma^2$ and compute the covariance for the model **x**

$$\mathrm{Cov}(\mathbf{x}_{L2}) = \sigma^2 \mathbf{V}\mathbf{S}^{-2}\mathbf{V}^{\mathrm{T}}. \tag{5.18}$$

The corresponding 95% confidence interval for **x** is

$$\Delta\mathbf{x}_{0.95} = 1.96 \times \sqrt{\mathrm{diag}(\mathrm{Cov}(\mathbf{x}_{L2}))} \tag{5.19}$$

and the model is recovered by inversion as

$$\mathbf{x} = \mathbf{x}_{L2} \pm \Delta\mathbf{x}_{0.95} \tag{5.20}$$

where $\mathbf{x}_{L2}$ is the least squares solution and the standard deviation for **x** is $\sigma_m = 0.5 \times \Delta\mathbf{x}_{0.95}$.

Figure 5.20 shows the 95% confidence interval $\Delta\mathbf{x}_{0.95}$ versus the model layer number $j$ in the matrix **A** discretized with different numbers of the model layers $J$ and computed assuming an experimental error of $\sigma = 1$ nV. A high sensitivity of shallow layers to the error is caused by an insufficient number of pulses with small pulse moments. Deep layers have the smallest singular values which explains their high sensitivity to the error.

The 2D and 3D implementations use Tx and Rx loops installed on the surface around investigated geological formation. Loop areas could be overlapped or side by side and their number depends on the formation geometry (figure 5.21).

The complete measuring procedure comprises measurements of the MRS signals versus pulse moment using one loop as the Tx loop and all others as the Rx loops. Each loop is assigned once as the Tx loop and always acts as the Rx loop. All these measurements are processed as one 2D or 3D data set. This procedure allows the maximum resolution. However, MRS loops are made of thick copper wire

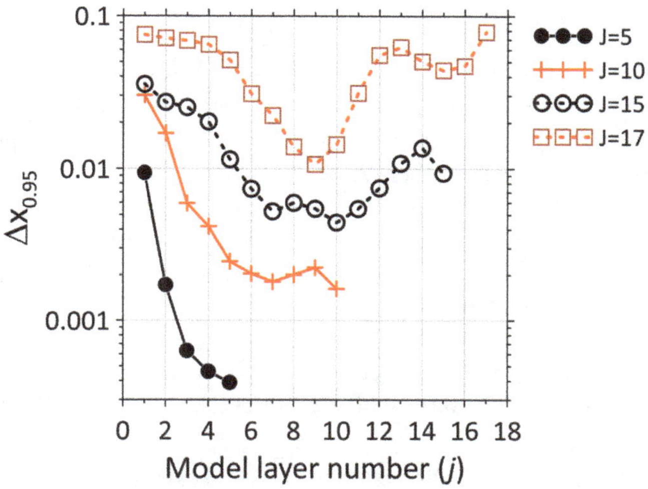


**Figure 5.20.** The 95% confidence interval versus the model layer number $j$ computed assuming discretization with different numbers of model layers $J$ in the matrix **A** (5, 10, 15, 17).

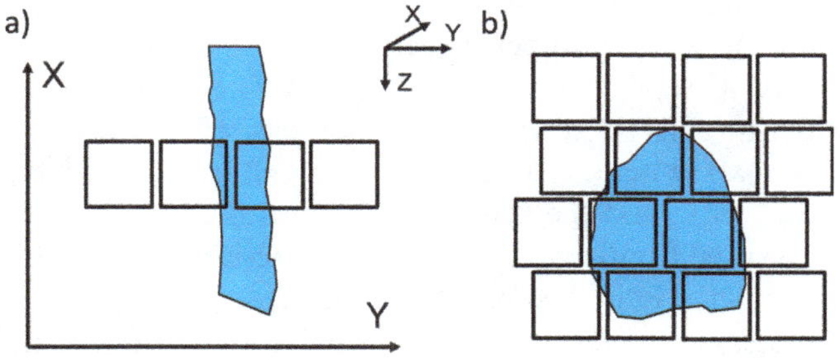


**Figure 5.21.** MRS loops around investigated geological formation: (a) 2D setup; (b) 3D setup.

supporting large current (40-ms-pulses of a current up to 600 A) with a high voltage insulation (up to 4 kV). They are heavy and expensive. Installation of each loop takes at least 20 min and may be longer in difficult field conditions. The complete imaging procedure comprises using many loops, which renders the instrument bigger and more expensive. Sometimes, measuring conditions do not allow for fulfillment of the entire 3D measurement in one attempt, which may require picking up and then reinstalling the loops.

The complete imaging procedure can be simplified by using the compact procedure which requires only one coincident Tx/Rx loop and comprises measuring the MRS signal in this loop while varying the pulse moment (standard 1D sounding). Thus, the 3D procedure is cut into independent soundings. For each sounding, the loop is set on the position assigned after the 2D or 3D imaging scheme (figure 5.21). All individual soundings compose a data set for 2D or 3D inversion.

Similar to the 1D MRS inverse problem, the 2D and 3D inversions require solving equation (5.8). For that, the integral equations corresponding to the method of measuring have to be discretized. In the 3D case, $\mathbf{A} = \left[\tilde{a}_{i,j}\right]$ is a rectangular matrix of $I \times J$ ($i = 1,2,..., I$; $j = 1,2,..., J$; $I = K \times L$; $J = J_X \times J_Y \times J_Z$; $K = \sum_{l=1}^{L} K_l$; $k_l = 1,2,..., K_l$; $l = 1,2,..., L$; $j_X = 1,2,..., J_X$, $j_Y = 1,2,..., J_Y$ and $j_Z = 1,2,..., J_Z$), $K_l$ is the number of pulse moments for a loop $l$, $L$ is the number of measuring loops, $J_X$, $J_Y$, $J_Z$ are the number of cells in the $X$, $Y$, and $Z$ directions, respectively. In the 2D case, either $J_X = 1$ or $J_Y = 1$. The elements of matrix **A** represent the amplitude of the magnetic resonance signal computed considering cells filled with bulk water (assuming a water content equal to one). The set of experimental data are $\mathbf{b} = (b_1, b_2, .., b_i, .., b_I)^T$, the model is $\mathbf{x} = (x_1, x_2, .., x_j, .., x_J)^T$, and the symbol $T$ denotes the transposition. For the 3D inversion, the subsurface has to be represented by model cells instead of the model layers used in the 1D case. These cells should be selected by varying their size and location below the measuring setup so that the model resolution matrix of the matrix **A** is close to the identity matrix ($\mathbf{R}_m = \mathbf{I}$).

For the time-lapse inversion, $\mathbf{A} = \left[\tilde{a}_{i,j}\right]$ is a rectangular matrix of $I \times J$ ($i = 1,2,..., I$; $j = 1,2,..., J$; $I = K \times L$; $J = J_t \times J_Z$; $K = \sum_{l=1}^{L} K_l$; $k_l = 1,2,..., K_l$; $l = 1,2,..., L$; $j_t = 1,2,..., J_t$ and $j_Z = 1,2,..., J_Z$), $K_l$ is the number of pulse moments for a loop $l$, $L$ is the number of measuring loops, $J_t$ is the number of time steps in the inverse model and $J_Z$ is the number of cells in the $Z$ directions, respectively. The elements of matrix **A** represent the amplitude of the magnetic resonance signal computed considering cells filled with bulk water (assuming a water content equal to one). The set of experimental data are $\mathbf{b} = (b_1, b_2, .., b_i, .., b_I)^T$, the model is $\mathbf{x} = (x_1, x_2, .., x_j, .., x_J)^T$, and the symbol $T$ denotes the transposition.

The SVD can be used for discretization of the time-lapse, 2D and 3D inversions in the same way as for the 1D inversions. The time-lapse, 2D and 3D MRS inversions have more parameters to define empirically or using assumptions, and require much more computational resources. However, the approach we used for the 1D inversion can be also applied to the time-lapse, 2D and 3D inversions. For that reason, and because of rare reported applications of the MRS 2D and 3D in hydrogeology, we limit the detailed analysis to the 1D case.

In this chapter, we have shown that the accuracy and stability of the solution of the MRS discrete inverse problem depends on measuring conditions and discretization of the initial integral equation. Discretization carried out with a smaller number of layers renders inversion more stable but not necessarily more accurate because the rough discretization may increase the discrepancy between the geometry of investigated geological formations and their numerical representation by the matrix **A**. Discretization of the matrix **A** imposes geometry of the solution but does not resolve the equivalence problem. The geometry constrain only reduces the number of equivalent models in the solution.

## 5.3 Inversion procedure

Whatever was discretization, the MRS inverse problem with real data does not allow for getting the exact solution corresponding precisely to real geological formations. The exact solution may exist only for synthetic data. However, MRS inversion provides an approximate inverse model that reproduces input data with a good accuracy and this inverse model is stable. Discrepancy between experimental data and data computed after the inverse model is estimated using the root-mean-square error (RMSE).

$$\text{RMSE} = \sqrt{\frac{1}{I}\sum_{i=1}^{I}\left(\sum_{j=1}^{J}a_{i,j}x_j - b_i\right)^2}. \tag{5.21}$$

Inversion stability should be verified in terms of a low sensitivity of the solution to small changes in input data. For stabilizing inversion, one should use justified assumptions or prior information about the solution. One of the popular methods for solving the MRS inverse problem is the Tikhonov regularization method. This method assumes the smoothness of the solution and comprises minimizing the Tikhonov functional $TF(\eta)$

$$TF(\eta) = \|\mathbf{Ax} - \mathbf{b}\|_{L_2} + \eta\,\|\mathbf{x}\|_{L_2} = \min \tag{5.22}$$

where $\eta(\sigma) > 0$ is called the smoothing factor. If $\sigma \to 0$ then $\eta \to 0$ and the approximate solution converges to the exact solution if the exact solution exists.

The Tikhonov regularization requires fitting experimental data with the discrepancy corresponding to an experimental error $\delta$

$$\|\mathbf{Ax} - \mathbf{b}\|_{L_2} \leqslant \delta. \tag{5.23}$$

The second term in equation (5.22) represents the smoothness assumption ($\eta\,\|\mathbf{x}\|_{L_2}$). While equation (5.23) has to be always respected, selection of the smoothing factor $\eta$ is more subjective and depends on prior information about the solution and/or assumptions. The Tikhonov functional $TF(\eta)$ has only one minimum which corresponds to the trade-off between the accuracy of fitting experimental data and the solution smoothness. It is convenient to select the solution corresponding to the minimum of $TF(\eta)$. For each value of $\eta$, we solve equation (5.22) using the conjugate gradient method and assuming a non-negative solution justified by the physical meaning of the water content and the relaxation times. Figure 5.22 shows an example of the Tikhonov functional computed versus the smoothing factor.

Regularization damps the effect of small singular values which cause inversion instability and acts as a low-pass filter for the solution. The Tikhonov filter function $\mathbf{F}$ is a diagonal matrix with the diagonal elements

$$f_j = \frac{s_j^2}{s_j^2 + \eta} \tag{5.24}$$

where $s_j$ are the singular values of the matrix $\mathbf{A}$.

The filter factor $\mathbf{F}$ damps scattering of the diagonal elements of the model resolution matrix $\mathbf{R}_{m,\,\eta}$

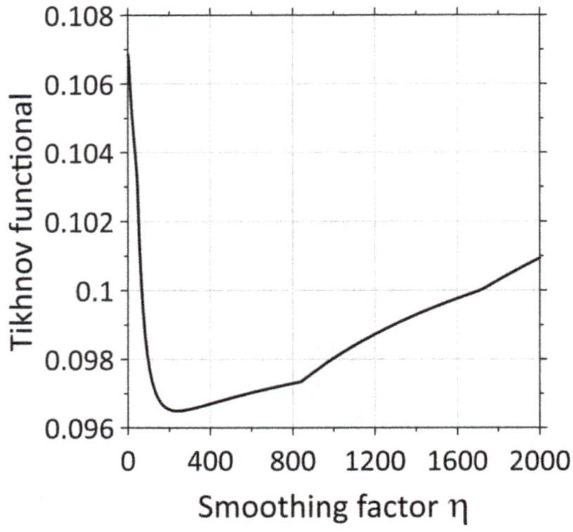


**Figure 5.22.** An example of the Tikhonov functional computed versus the smoothing factor.

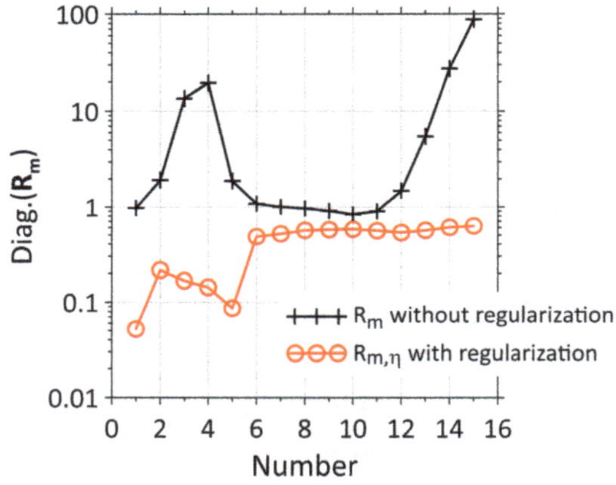


**Figure 5.23.** The diagonal elements of the model resolution matrix without regularization $\mathbf{R}_m$ (black crosses) and with regularization $\mathbf{R}_{m,\eta}$ (red circles).

$$\mathbf{R}_{m,\,\eta} = \mathbf{VFV}^{\mathrm{T}}. \tag{5.25}$$

Let us consider an example of the matrix $\mathbf{A}$ discretized with seventeen layers ($J = 17$) so that $\mathbf{R}_m \neq \mathbf{I}$. Figure 5.23 shows the diagonal elements of the model resolution matrix without regularization (black crosses, $\eta = 0$) and with regularization (red circles, $\eta > 0$). Without regularization, the model will be recovered with scattered errors corresponding to fluctuations of the diagonal elements of the model resolution matrix. Regularization damps scattering of the diagonal elements but does not improve the resolution because of $\mathbf{R}_{m,\,\eta} \neq \mathbf{I}$.

If $\eta > 0$ then the solution covariance and the 95% confidence interval for the regularized solution $\Delta\mathbf{x}_\eta$ becomes

$$\mathrm{Cov}(\mathbf{x}_\eta) = \sigma^2\mathbf{VFS}^{-2}\mathbf{V}^{\mathrm{T}} \tag{5.26}$$

$$\Delta \mathbf{x}_\eta = 1.96 \times \sqrt{\mathrm{diag}(\mathrm{Cov}(\mathbf{x}_\eta))} \tag{5.27}$$

$$\mathbf{x} = \mathbf{x}_{L2} \pm \Delta \mathbf{x}_\eta. \tag{5.28}$$

Note that the confidence interval $\Delta \mathbf{x}_\eta$ estimates the uncertainty for the regularized solution and not the uncertainty for the true model and if $\eta > 0$ then $\Delta \mathbf{x}_{0.95} \geqslant \Delta \mathbf{x}_\eta$.

In the following example, we discretize the matrix **A** with fourteen layers ($J = 14$) and the model resolution matrix $\mathbf{R}_m = \mathbf{I}$ as shown in figure 5.24(a) by black crosses. However, the recovered model is sensitive to the data errors estimated with the 95% confidence interval $\Delta \mathbf{x}_{0.95}$ computed assuming the data error $\sigma = 1$ nV (figure 5.24(b), black crosses). Regularization reduces the uncertainty $\Delta \mathbf{x}_\eta$ (figure 5.24(b), red circles) but causes $\mathbf{R}_{m,\eta} \neq \mathbf{I}$ as shown in figure 5.24(a) by red circles. Thus, regularization stabilizes inversion by damping oscillations in the recovered inverse model but degrades the resolution.

The SVD analysis provides an estimate of the 95% confidence interval for the model parameters computed assuming i.i.d. normal data errors. The 95% confidence interval is the property of the matrix **A** and is independent of the model. However, the data errors may have other than normal distribution and one may be required to estimate the uncertainty in the specific model. For that, it is convenient to use the Monte Carlo simulations. One can use any inverse model that fits experimental data within experimental errors and apply to this model random perturbations $u_{n,j}$. We apply the Monte Carlo analysis to inversion of FID measurements for the water content and neglect the relaxation ($\mathbf{T} = \mathbf{I}$ in equation (5.3)). The water content $w_{n,j}$ in the layer $j$ ($j = 1,2,.., J$) of the perturbed model $n$ ($n = 1,2,.., N$) becomes

$$\begin{cases} w_{n,j} = w_j + \Delta w_{n,j} \times u_{n,j} \\ \text{if } \; w_{n,j} < 0 \; \text{ then } w_{n,j} = 0 \end{cases} \tag{5.29}$$

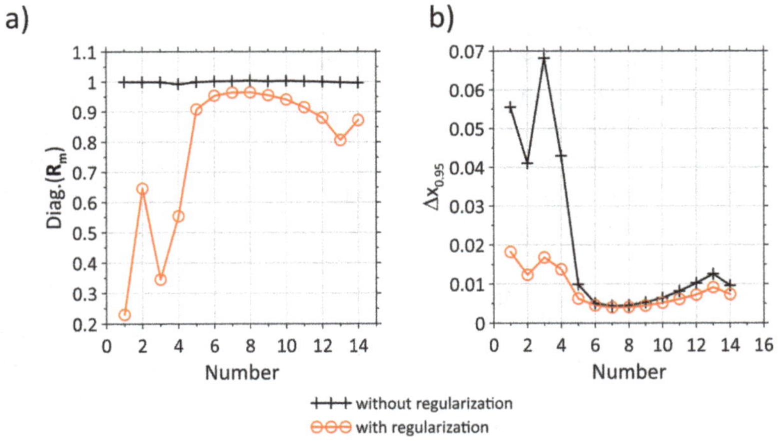


**Figure 5.24.** (a) The diagonal elements of the model resolution matrix without regularization (black crosses) and with regularization (red circles). (b) The corresponding 95% confidence intervals.

where $w_j$ is the water content model provided by the Tikhonov regularization, $\Delta w_{nj}$ is the 95% confidence interval for the regularized solution (layer $j$) and $u_{n,j}$ are the random numbers.

The multiply-with-carry method (MWC) allows the generating of pseudo-random numbers uniformly distributed between −1 and 1. All perturbed models respect the 95% confidence interval estimated with the SVD, but some of them may not fit experimental data. The fitting error is estimated by computing the root-mean-square error $\text{RMSE}_n$ (equation (5.21)). The model perturbations are random and hence values of $\text{RMSE}_n$ are also random numbers characterized by the mean $\mu$ and the standard deviation $\sigma$

$$\begin{cases} \mu = \dfrac{1}{N}\displaystyle\sum_{n=1}^{N}\text{RMSE}_n \\ \sigma = \sqrt{\dfrac{1}{N}\displaystyle\sum_{n=1}^{N}(\text{RMSE}_n - \mu)^2} \end{cases}. \tag{5.30}$$

The probability density function (PDF) is a function whose value at any point provides a relative likelihood that the value of the random variable $\text{RMSE}_n$ would equal that point. The PDF for the normally distributed random values can be computed using the mean $\mu$ and the standard deviation $\sigma$

$$\text{PDF}_n = \frac{1}{\sigma\sqrt{2\pi}}\exp\left(-\frac{(\text{RMSE}_n - \mu)^2}{2\sigma^2}\right). \tag{5.31}$$

For a set of experimental data with unknown distribution, the PDF can be directly computed by counting the number of perturbed models with $\text{RMSE}_n$ corresponding to each of $M$ equal intervals ($m = 1,2,.., M$)

$$\Delta\text{RMSE}_m = \text{M}^{-1}(\text{RMSE}_{\text{max}} - \text{RMSE}_{\text{min}}). \tag{5.32}$$

Perturbation of the regularized inverse model provides solutions that fit experimental data with different accuracy. The uncertainty can be estimated by computing the water column $H$ (equation (5.7)) for models that fit experimental data with equal RMSE (the equivalent models) and selecting two extreme equivalent models corresponding to the maximum $H_{\text{max}}$ and the minimum $H_{\text{min}}$ values of the water column. Selection of the equivalent solutions for computing the water column has to be done using the regularized solution ($\text{RMSE}_\eta$) or the experimental error $\delta$ estimated making the noise measurements. Figure 5.25 shows an example of the Monte Carlo simulation. The equivalent water column computed for $10^5$ perturbed models is plotted versus the fitting error. For each value of the RMSE, the mean, the standard deviation and the 95% confidence interval of the equivalent water column estimate correspond to the uncertainty in the equivalent inverse models.

The 2D MRS inversion for the water content $\mathbf{w}$ comprises minimization of the Tikhonov functional

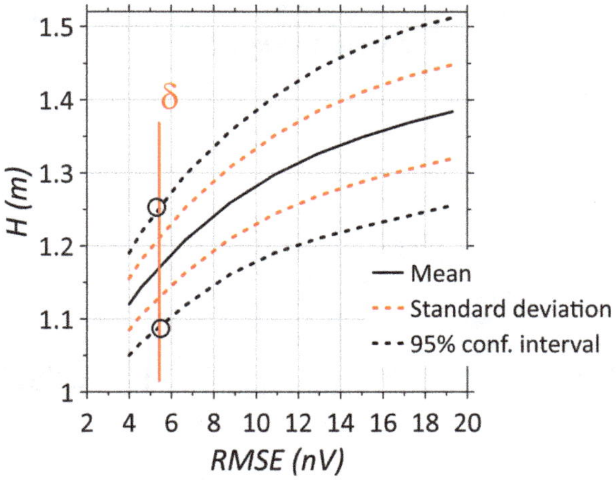


**Figure 5.25.** The water column $H$ versus RMSE computed using the $10^5$ perturbed model. The red vertical line is set equal to the experimental errors and the black circles show the maximum and the minimum values of the water column corresponding to two equivalent extreme models.

$$\mathrm{TF}_{2D}(\eta) = \left\| \mathbf{A}\mathbf{w}_\eta - \mathbf{v}_1 \right\|_{L_2} + \eta_X \times \left( \frac{\partial}{\partial X}\mathbf{w}_\eta \right)^2 + \eta_Z \times \left( \frac{\partial}{\partial Z}\mathbf{w}_\eta \right)^2 = \min \tag{5.33}$$

where $X$ is the coordinate of the measuring loops and $Z$ is the depth. Selection of the smoothing factors $\eta_X$ and $\eta_Z$ depends on the anisotropy of the subsurface. The 3D MRS inversion comprises minimization of the Tikhonov functional

$$\begin{aligned}\mathrm{TF}_{3D}(\eta) = \left\| \mathbf{A}\mathbf{w}_\eta - \mathbf{v}_1 \right\|_{L_2} + \eta_X \times \left( \frac{\partial}{\partial X}\mathbf{w}_\eta \right)^2 \\ + \eta_Y \times \left( \frac{\partial}{\partial Y}\mathbf{w}_\eta \right)^2 + \eta_Z \times \left( \frac{\partial}{\partial Z}\mathbf{w}_\eta \right)^2 = \min.\end{aligned} \tag{5.34}$$

The time-lapse MRS inversion requires 1D time-lapse measurements with the measuring loop at the same place, but it can be extended to the 2D and 3D studies. The 1D time-lapse inversion comprises minimization of the Tikhonov functional

$$\mathrm{TF}_t(\eta) = \left\| \mathbf{A}\mathbf{w}_\eta - \mathbf{v}_1 \right\|_{L_2} + \eta_t \times \left( \frac{\partial}{\partial t}\mathbf{w}_n \right)^2 + \eta_Z \times \left( \frac{\partial}{\partial Z}\mathbf{w}_\eta \right)^2 = \min. \tag{5.35}$$

The general approach for the time-lapse, 2D and 3D MRS inversions is like the 1D inversion and differs mainly by the size of data to process. In this book, we omit the detailed study of these inversions.

## Bibliography

Aster R C, Borchers B and Thurber C H 2005 *Parameter Estimation and Inverse Problems* (Amsterdam: Elsevier)

# Chapter 6

## Examples of 1D MRS inversion

For getting started, we show a typical procedure of the 1D magnetic resonance sounding (MRS) inversion for the water content. The 2D and 3D inversions and the inversion for the relaxation times is based on the similar procedure. Note that this procedure is not exclusive and other inversion methods can be used. For any inversion technique, the input data must be fitted with the accuracy equal or better than the experimental error. Indeed, if we would not respect experimental data, then we do not need field measurements.

In the following examples, we use the spike models discussed in the previous chapters (figures 5.18 and 5.19) and then we show an inversion of a real data set. For each example, we build different 1D inverse models by changing discretization of the integral equation and the regularization. The data sets are processed with the discretization done in three different ways: (1) the minimum number of layers that allow fitting the input data ($\mathbf{R}_m = \mathbf{I}$); (2) the maximum number of layers considering the model resolution matrix ($\mathbf{R}_m = \mathbf{I}$); (3) with 40 layers corresponding to the smooth inversion ($\mathbf{R}_m \neq \mathbf{I}$). Inversion is done by applying the Tikhonov regularization method. For this exercise, we use inversion of the amplitudes, which is less accurate but more robust than inversion of the complex signals.

We assume a 100-m-side square loop, a 100-Ω.m-half-space and the maximum pulse moment equal to 16 000 A-ms. We consider two one-layer models: the shallow water model and the deep water model. In these examples, the boundaries do not coincide with the model layers given by the discretization thus adding an inaccuracy even for noiseless data.

For investigating stability of the inverse model, we use the Monte Carlo approach. Analysis of the inversion results can be carried out in different ways. One may disturb the input data and analyze the impact of these perturbations on the inverse model. Otherwise, the inverse model can be disturbed. For that, the water content in some layers in the inverse model can be fixed using a prior knowledge about the subsurface or all the layers can be set free. The magnitude of the perturbations can be done using estimates of the experimental noise. In our examples, we make inversions of the noiseless, and noise contaminated input

doi:10.1088/978-0-7503-3155-5ch6

data. Then we apply to obtained inverse models random perturbations with the constant amplitude $\Delta w_{n,j} = 0.01$ for any $n$ and $j$ in equation (5.29). The result comprises selecting two equivalent models by computing the minimum ($H_{\min}$) and the maximum ($H_{\max}$) water column solutions out of $10^6$ disturbed models.

## 6.1 Noiseless synthetic data

MRS inversion depends on the accuracy of input data and we will use first the noiseless data sets. The resolution is always degraded when using noise contaminated experimental data but these examples show the theoretical limits for the MRS inversion.

### 6.1.1 Model: shallow water

The shallow water model comprises one layer between 12.42 and 16.84 m with the water content $w = 0.1$ and with $w = 0$ outside this layer.

*6.1.1.1 Smooth inversion with a 40-layer model*

The smooth inversion comprises many model layers in the matrix **A** and the SVD analysis shows that these layers cannot be resolved by the inversion because of $\mathbf{R}_m \neq \mathbf{I}$. For this reason, the smooth inversion always assumes regularization. The smoothing factor $\eta(\sigma) > 0$ is selected by minimizing the Tikhonov functional $\mathrm{TF}(\eta)$ in equation (5.22). Figure 6.1(a) shows the shallow water model (red line) and the regularized inverse model (black line). For investigating stability of this inverse model, we apply random perturbations. Figure 6.1(b) shows the regularized inverse model (red line) and the perturbed models corresponding to the $H_{\min}$ (black line) and the $H_{\max}$ (blue line) equivalent solutions. One can see that all the inverse models show the

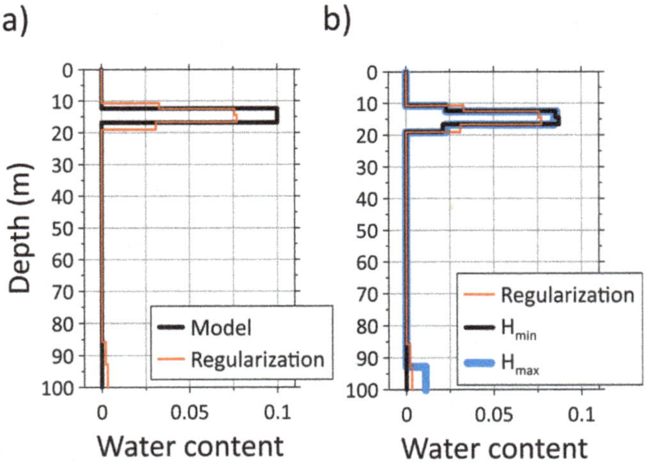


**Figure 6.1.** Smooth inversion of the noiseless synthetic data set with a 40-layer inverse model: (a) the true model (black line) and the regularized inverse model (red line); (b) the regularized inverse model (red line) and the perturbed equivalent inverse models corresponding to the minimum (black line) and maximum (blue line) water columns solutions.

layer between 10 and 20 m and hence, the shallow layer cannot be neglected. One of the equivalent solutions ($H_{min}$) shows that the input data can be fitted without the deep layer below 85 m thus suggesting that this layer may be an inversion artifact.

Three models are the equivalent models because the theoretical signals fit the input data with equal accuracy (RMSE $\approx$ 0.9nV). Figure 6.2 shows the input data and the theoretical fits corresponding to the equivalent models. Visually, the amplitudes computed with these equivalent models can be barely distinguished.

Figure 6.3(a) shows the probability density functions (PDF) computed using the perturbed inverse models (crosses) computed assuming the normal distribution and using equation (5.31). One can see that both distributions are similar and hence the RMSE is normally distributed. With our perturbations, the maximum probability for the RMSE is about 22 nV. It is much larger than the RMSE corresponding to the regularized solution. The probability of getting the RMSE less than 8 nV is very low, which shows that the inversion is very sensitive to variations in the input data. Figure 6.3(b) shows the mean of the water column ($H$) estimate versus RMSE with

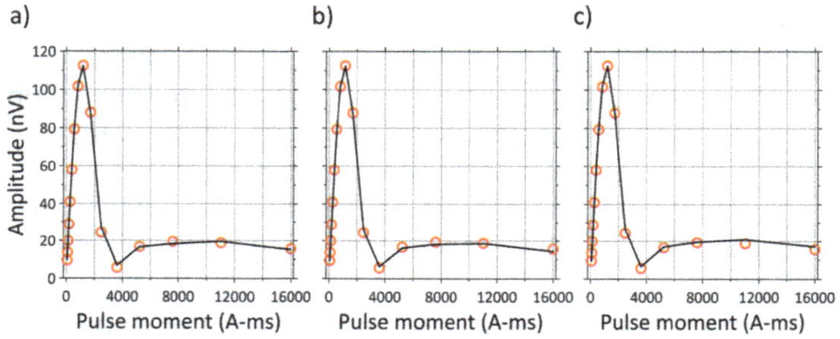


**Figure 6.2.** Smooth inversion of the noiseless synthetic data set with a 40-layer inverse model. The amplitude of a synthetic noiseless signal (red circles) fitted by the theoretical signal (black line) corresponding to the equivalent inverse models: (a) regularized inversion; (b) the minimum water column inversion; (c) the maximum water column inversion.

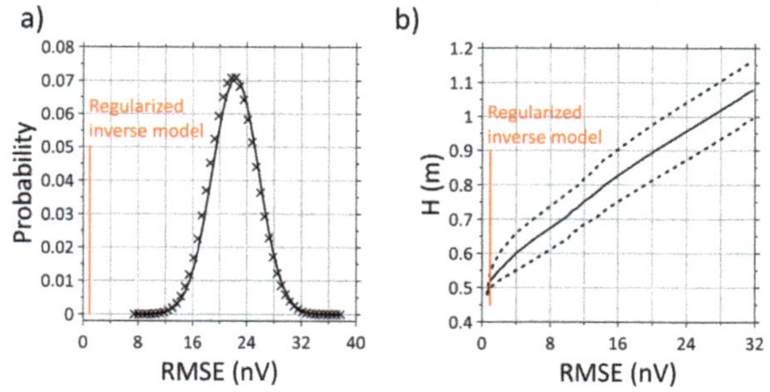


**Figure 6.3.** Smooth inversion of the noiseless synthetic data set with a 40-layer inverse model: (a) the probability density function versus RMSE computed directly (black crosses) and computed assuming the normal distribution using the mean and the standard deviation (solid black line); (b) the water column (black solid line) and the 95% confidence interval (black dashed lines) versus RMSE. The red vertical lines show the water columns corresponding to the inverse models that are equivalent to the regularized inverse model.

the 95% confidence interval (dashed lines). The red vertical line shows the water columns corresponding to the inverse models that are equivalent to the regularized inverse model.

*6.1.1.2 Inversion with the maximum resolution using a 17-layer model*

The maximum theoretical resolution corresponds to the maximum number of the model layers in the matrix **A** limited by the condition $\mathbf{R}_m = \mathbf{I}$. In our example shown in figures 6.4–6.6 the maximum number of model layers is 17. When using real data, experimental errors degrade the inversion stability and the maximum theoretical resolution is unachievable. Thus, inversion with 17 layers also requires regularization. Note that discretization with respect to the condition $\mathbf{R}_m = \mathbf{I}$ allows using the SVD for investigating properties of the matrix **A**. We compute three

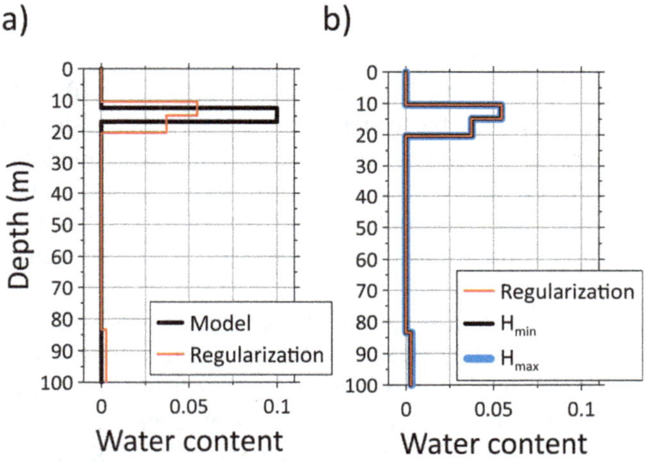


**Figure 6.4.** Inversion of the noiseless synthetic data set with a 17-layer inverse model: (a) the true model (black line) and the regularized inverse model (red line); (b) the regularized inverse model (red line) and the perturbed equivalent inverse models corresponding to the minimum (black line) and maximum (blue line) water columns solutions.

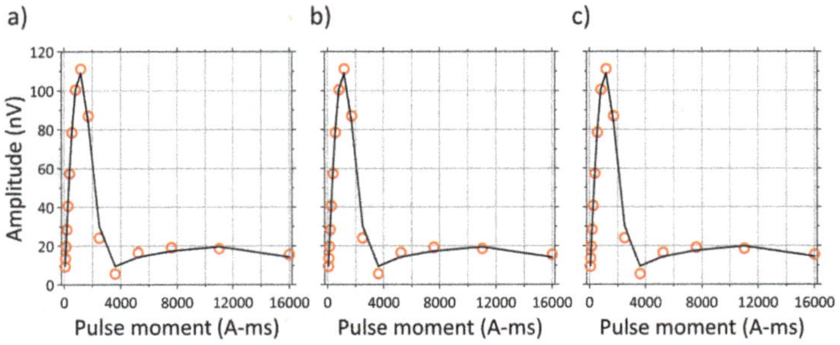


**Figure 6.5.** Inversion of the noiseless synthetic data set with a 17-layer inverse model. The amplitude of a synthetic noiseless signal (red circles) fitted by the theoretical signal (black line) corresponding to the equivalent inverse models: (a) regularized inversion; (b) the minimum water column inversion; (c) the maximum water column inversion.

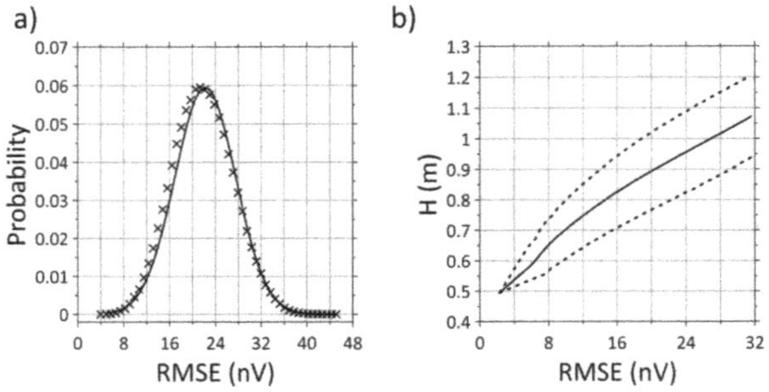


**Figure 6.6.** Inversion of the noiseless synthetic data set with a 17-layer inverse model: (a) the probability density function versus RMSE computed directly (black crosses) and computed assuming the normal distribution using the mean and the standard deviation (solid black line); (b) the water column (black solid line) and the 95% confidence interval (black dashed lines) versus RMSE.

equivalent inverse models that fit the input data with the RMSE ≈ 2.3nV. The inverse models are almost identical, which shows that inversion is stable. Inversion results suggest a thin deep layer with the water content 100 folds smaller than the shallow layer. This deep layer is an artifact explained by not coinciding boundaries of the model layers in the matrix **A** and the boundaries of the model. The artifact layers may appear when the forward modeling is not fully compatible with the subsurface. For example, when the 1D inversion is carried out in a 3D geological environment.

*6.1.1.3 Inversion with the minimum number of layers using a 12-layer model*

The maximum stability of the inversion can be achieved by using the minimum number of model layers in the matrix **A** that allow fitting the input data with the accuracy limited by the experimental errors. For each data set, the minimum number of layers may vary. For this inversion, the condition $\mathbf{R}_m = \mathbf{I}$ is automatically respected and regularization can be unnecessary. Inversion of the data set corresponding to the shallow water model requires 12 layers that allow fitting the input data with the RMSE ≈ 3.3nV. Results are shown in figures 6.7–6.9.

As well as the 17-layer inverse model, the 12-layer inverse model shows the deep artifact layer. However, the water content in the deep layer is more than 100 folds smaller than in the shallow layer.

### 6.1.2 Model: deep water

The deep water model comprises one layer between 48.85 and 61.7 m with the water content $w = 0.1$ and with $w = 0$ outside this layer.

*6.1.2.1 Smooth inversion with a 40-layer model*

Figures 6.10–6.12 show results of the smooth inversion of the input data set corresponding to the deep water model. The model is well resolved and the input data are fitted with the RMSE ≈ 1.1nV.

*6.1.2.2 Inversion with the maximum resolution using a 17-layer model*

Figures 6.13–6.15 show results of inversion using a 17-layer model. Comparison of the deep water model and the inverse model shows that the depth of the model is well

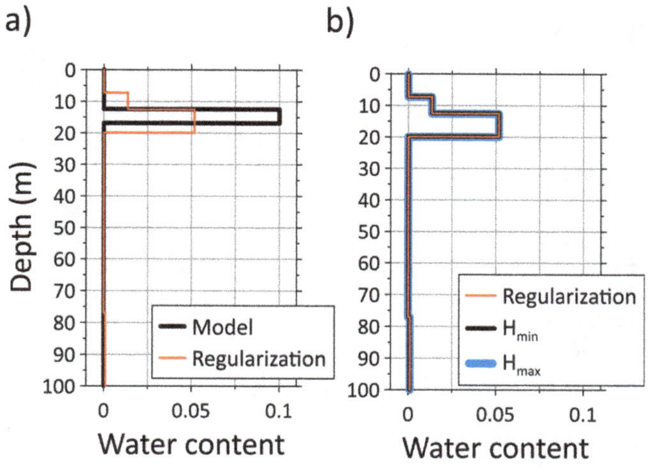


**Figure 6.7.** Inversion of the noiseless synthetic data set with a 12-layer inverse model: (a) the true model (black line) and the regularized inverse model (red line); (b) the regularized inverse model (red line) and the perturbed equivalent inverse models corresponding to the minimum (black line) and maximum (blue line) water columns solutions.

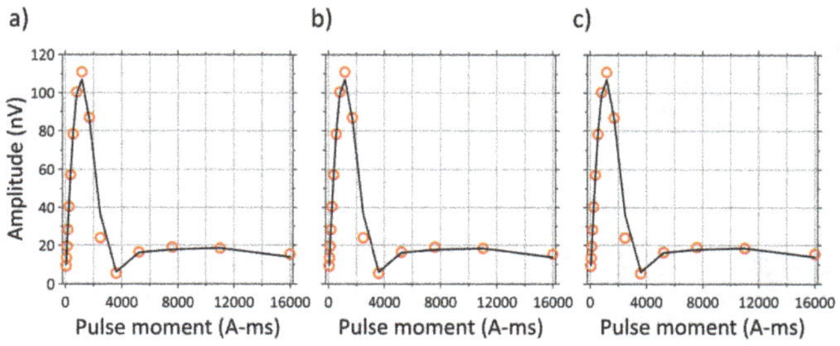


**Figure 6.8.** Inversion of the noiseless synthetic data set with a 12-layer inverse model. The amplitude of a synthetic noiseless signal (red circles) fitted by the theoretical signal (black line) corresponding to the equivalent inverse models: (a) regularized inversion; (b) the minimum water column inversion; (c) the maximum water column inversion.

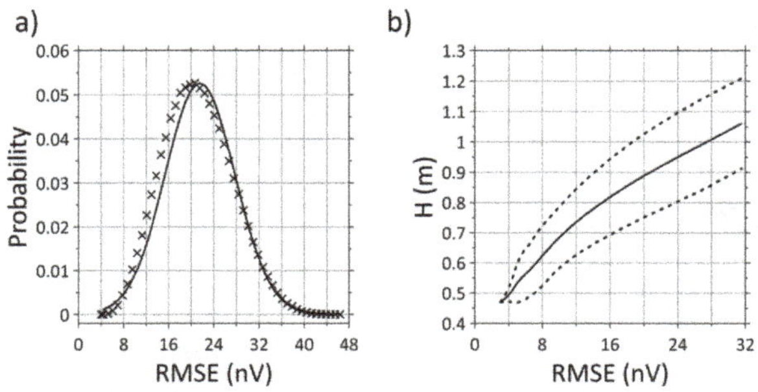


**Figure 6.9.** Inversion of the noiseless synthetic data set with a 12-layer inverse model: (a) the probability density function versus RMSE computed directly (black crosses) and computed assuming the normal distribution using the mean and the standard deviation (solid black line); (b) the water column (black solid line) and the 95% confidence interval (black dashed lines) versus RMSE.

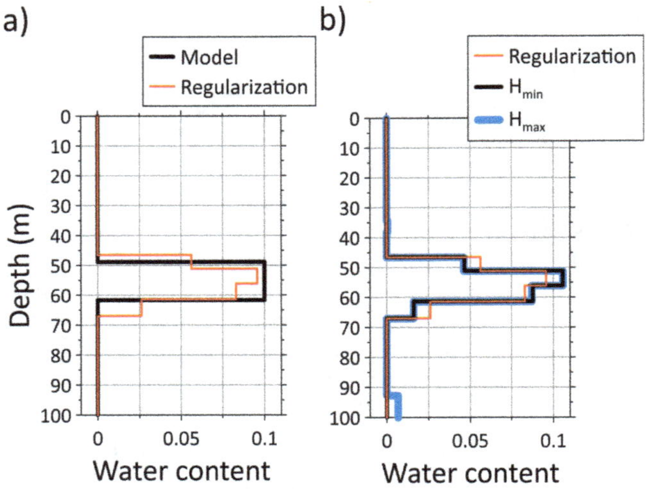


**Figure 6.10.** Smooth inversion of the noiseless synthetic data set with a 40-layer inverse model: (a) the true model (black line) and the regularized inverse model (red line); (b) the regularized inverse model (red line) and the perturbed equivalent inverse models corresponding to the minimum (black line) and maximum (blue line) water columns solutions.

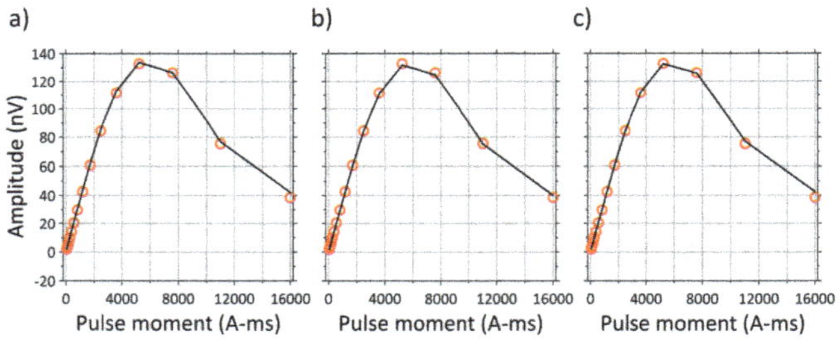


**Figure 6.11.** Smooth inversion of the noiseless synthetic data set with a 40-layer inverse model. The amplitude of a synthetic noiseless signal (red circles) fitted by the theoretical signal (black line) corresponding to the equivalent inverse models: (a) regularized inversion; (b) the minimum water column inversion; (c) the maximum water column inversion.

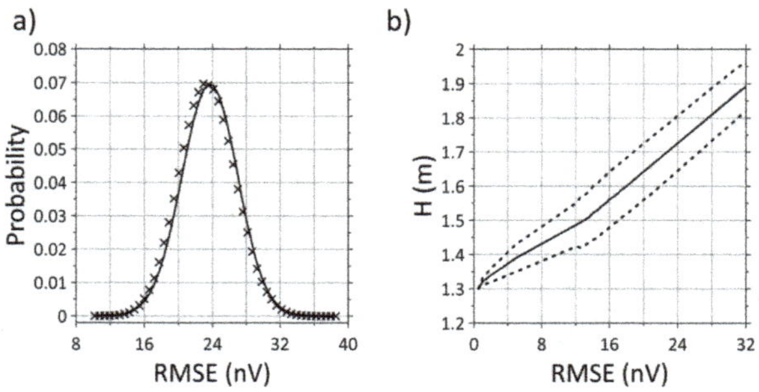


**Figure 6.12.** Inversion of the noiseless synthetic data set with a 40-layer inverse model: (a) the probability density function versus RMSE computed directly (black crosses) and computed assuming the normal distribution using the mean and the standard deviation (solid black line); (b) the water column (black solid line) and the 95% confidence interval (black dashed lines) versus RMSE.

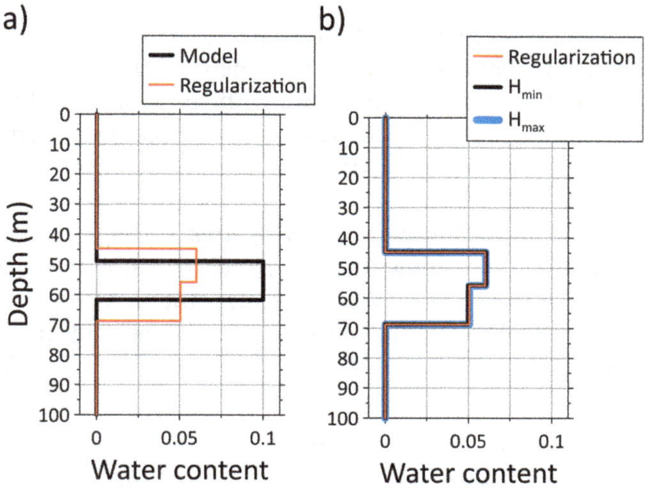


**Figure 6.13.** Inversion of the noiseless synthetic data set with a 17-layer inverse model: (a) the true model (black line) and the regularized inverse model (red line); (b) the regularized inverse model (red line) and the perturbed equivalent inverse models corresponding to the minimum (black line) and maximum (blue line) water columns solutions.

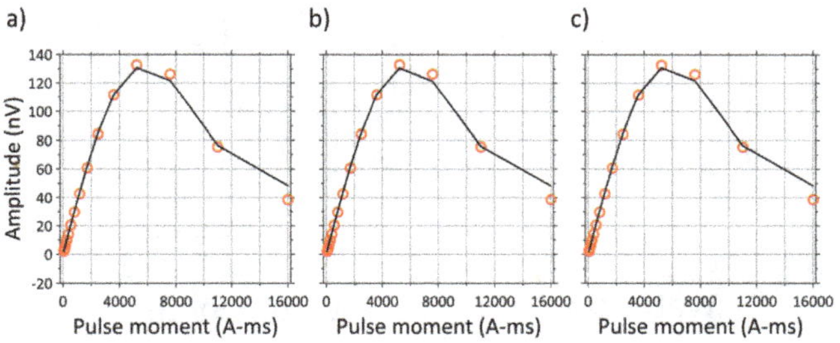


**Figure 6.14.** Inversion of the noiseless synthetic data set with a 17-layer inverse model. The amplitude of a synthetic noiseless signal (red circles) fitted by the theoretical signal (black line) corresponding to the equivalent inverse models: (a) regularized inversion; (b) the minimum water column inversion; (c) the maximum water column inversion.

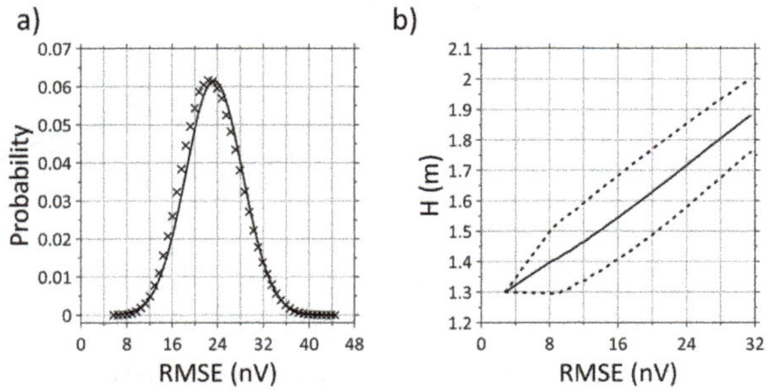


**Figure 6.15.** Inversion of the noiseless synthetic data set with a 17-layer inverse model: (a) the probability density function versus RMSE computed directly (black crosses) and computed assuming the normal distribution using the mean and the standard deviation (solid black line); (b) the water column (black solid line) and the 95% confidence interval (black dashed lines) versus RMSE.

recovered, but the water content is underestimated. This can be explained by the discrepancy between the deep water model geometry and the model layer boundaries. However, the input data are well fitted with the RMSE ≈ 2.8 nV showing a good accuracy of the inversion. These results reveal the equivalence between the deep water model and the inverse model.

### *6.1.2.3 Inversion with the minimum number of layers using an 8-layer model*

Eight model layers allow fitting the input data with the RMSE ≈ 4.4nV. Selecting a smaller number of layers degrades the RMSE. Inversion results reveal the equivalence between the deep water model and the inverse model (figures 6.16–6.18).

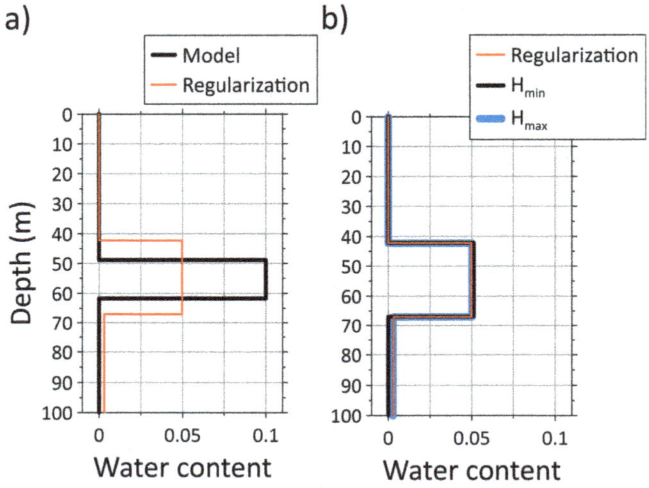


**Figure 6.16.** Inversion of the noiseless synthetic data set with an 8-layer inverse model: (a) the true model (black line) and the regularized inverse model (red line); (b) the regularized inverse model (red line) and the perturbed equivalent inverse models corresponding to the minimum (black line) and maximum (blue line) water columns solutions.

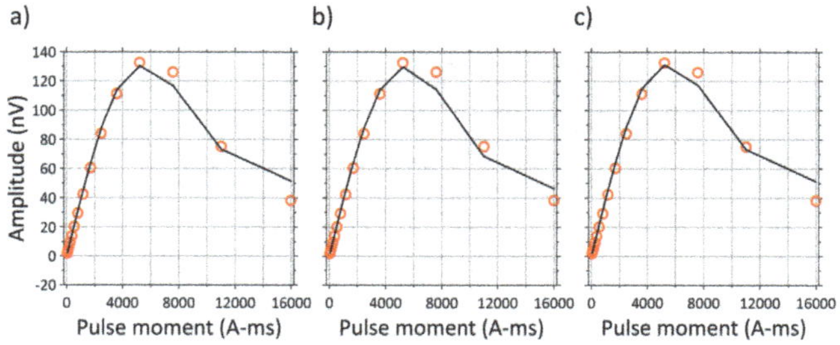


**Figure 6.17.** Inversion of the noiseless synthetic data set with an 8-layer inverse model. The amplitude of a synthetic noiseless signal (red circles) fitted by the theoretical signal (black line) corresponding to the equivalent inverse models: (a) regularized inversion; (b) the minimum water column inversion; (c) the maximum water column inversion.

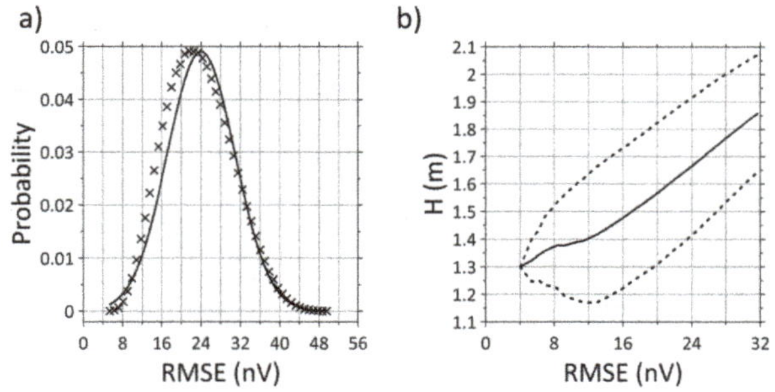


**Figure 6.18.** Inversion of the noiseless synthetic data set with an 8-layer inverse model: (a) the probability density function versus RMSE computed directly (black crosses) and computed assuming the normal distribution using the mean and the standard deviation (solid black line); (b) the water column (black solid line) and the 95% confidence interval (black dashed lines) versus RMSE.

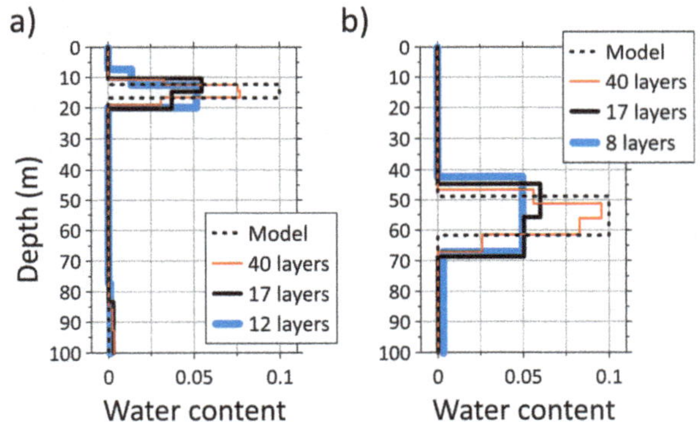


**Figure 6.19.** Inversion of the noiseless synthetic data sets: (a) the shallow water model of the water content (black dashed line) and three inverse models corresponding to the regularized solution using 40, 17 and 12 model layers; (b) the deep water model (black dashed line) and three inverse models corresponding to the regularized solution using 40, 17 and 8 model layers.

### 6.1.3 Summary

Inversion of the noiseless synthetic data sets corresponding to the shallow and the deep water models and carried out with different number of model layers shows that the models are recovered with better accuracy when using the 40-layer inversion. Figure 6.19 shows that the 17-, 12- and 8-layer inversions underestimate twice the water content and overestimate the thickness of the model. The main reason for these errors is the discretization of the matrix **A** with the boundaries of model layers not coinciding with the water model boundaries.

The results of the noiseless data inversion are summarized in table 6.1.

**Table 6.1.** The water column (*H*), the RMSE and the relative error of the water column estimates for the noiseless data sets.

| | Model: shallow water | | | |
|---|---|---|---|---|
| Number | Inverse model | RMSE (nV) | *H* (m) | Error *H* (%) |
| | True model | | 0.442 | |
| | ***Solution $H_{min}$*** | | | |
| 1 | 40 layers | 0.74 | 0.448 | 1.3 |
| 2 | 17 layers | 2.29 | 0.490 | 10.9 |
| 3 | 12 layers | 3.35 | 0.471 | 6.6 |
| | ***Regularization*** | | | |
| 4 | 40 layers | 0.95 | 0.487 | 10.2 |
| 5 | 17 layers | 2.29 | 0.493 | 11.6 |
| 6 | 12 layers | 3.35 | 0.475 | 7.5 |
| | ***Solution $H_{max}$*** | | | |
| 7 | 40 layers | 0.91 | 0.527 | 19.3 |
| 8 | 17 layers | 2.29 | 0.499 | 13.0 |
| 9 | 12 layers | 3.34 | 0.475 | 7.4 |
| | Model: deep water | | | |
| | Inverse model | RMSE (nV) | *H* (m) | Error *H* (%) |
| | True model | | 1.285 | |
| | ***Solution $H_{min}$*** | | | |
| 10 | 40 layers | 0.60 | 1.291 | 0.5 |
| 11 | 17 layers | 2.81 | 1.313 | 2.2 |
| 12 | 8 layers | 4.43 | 1.267 | −1.4 |
| | ***Regularization*** | | | |
| 13 | 40 layers | 1.19 | 1.317 | 2.5 |
| 14 | 17 layers | 2.81 | 1.317 | 2.5 |
| 15 | 8 layers | 4.44 | 1.337 | 4.0 |
| | ***Solution $H_{max}$*** | | | |
| 16 | 40 layers | 1.11 | 1.338 | 4.1 |
| 17 | 17 layers | 2.81 | 1.318 | 2.5 |
| 18 | 8 layers | 4.43 | 1.339 | 4.2 |

Figure 6.20 shows the relative error of the water column estimate versus the number of the inverse model in table 6.1. The error for the inverse models 1–9 is larger than that for the inverse models 10–18, thus showing that the water column is better estimated for the deep water model composed of a thick water saturated layer.

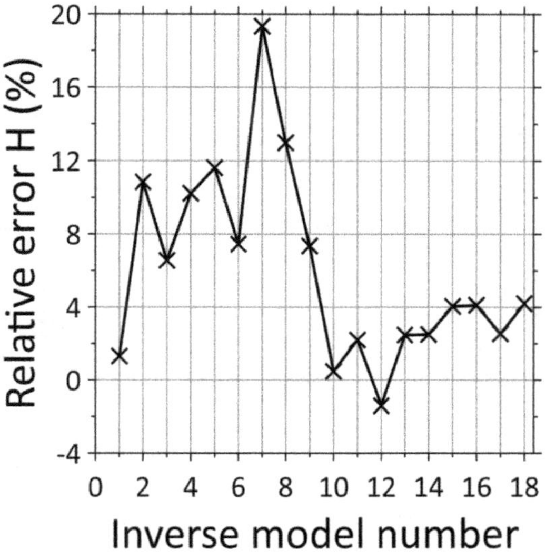


**Figure 6.20.** The relative error of the water column estimates for the noiseless data sets versus the inverse model number.

## 6.2 Noise contaminated synthetic data

The next exercise is carried out using the synthetic signal but contaminated with normally distributed random noise.

### 6.2.1 Model: shallow water

The noise standard deviation for the shallow water model $\sigma = 15.23\text{nV}$ which corresponds to 13.5% of the maximum amplitude of the MRS signal.

*6.2.1.1 Smooth inversion with a 40-layer model*

Figures 6.21–6.23 show the 40-layer inverse models. The regularized inversion shows an approximate location of the model layer with the water content significantly underestimated. The perturbed equivalent models also point to the shallow water location, but split the model into two layers. Two models of three show deep water caused by the noise influence.

The input data are well fitted by the inversion with the fitting error RMSE = 5.7nV.

The probability of getting the RMSE = 5.7nV is small showing that the great majority of the perturbed inverse models do not fit the input data with the accuracy equal to the regularized inverse model.

*6.2.1.2 Inversion with the maximum resolution using a 17-layer model*

Inversion with the 17-layer inverse model shows the location of the shallow water model and underestimates the water content. Inversion suggests the existence of deep water. This deep water solution can be explained by the noise influence (figure 6.24).

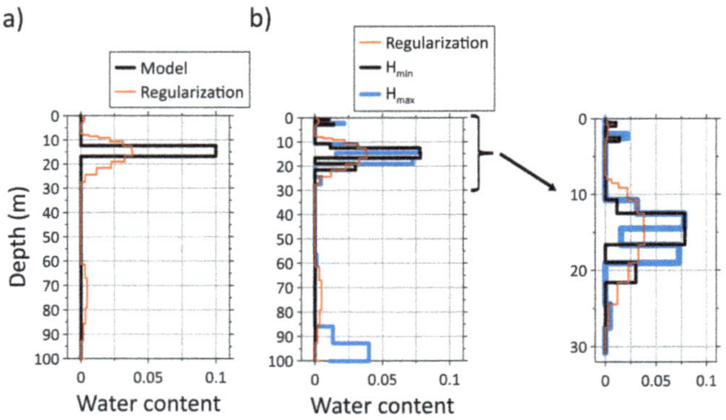


**Figure 6.21.** Smooth inversion of the noise contaminated synthetic data set with a 40-layer inverse model: (a) the true model (black line) and the regularized inverse model (red line); (b) the regularized inverse model (red line) and the perturbed equivalent inverse models corresponding to the minimum (black line) and maximum (blue line) water columns solutions. The right hand graph shows a zoom of the shallow part of the water content profile.

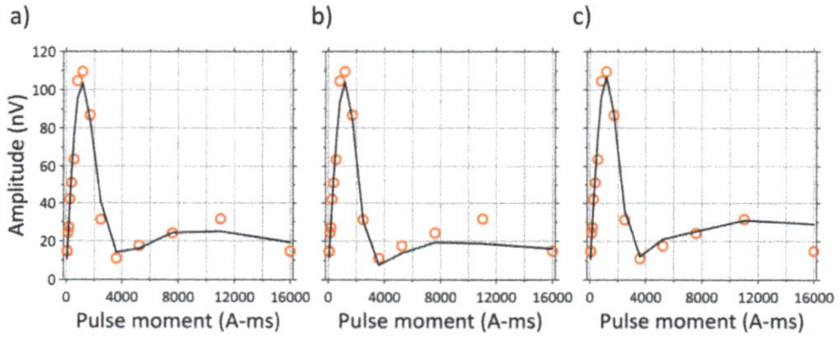


**Figure 6.22.** Smooth inversion of the noise contaminated synthetic data set with a 40-layer inverse model. The signal amplitude (red circles) fitted by the theoretical signal (black line) corresponding to the equivalent inverse models: (a) regularized inversion; (b) the minimum water column inversion; (c) the maximum water column inversion.

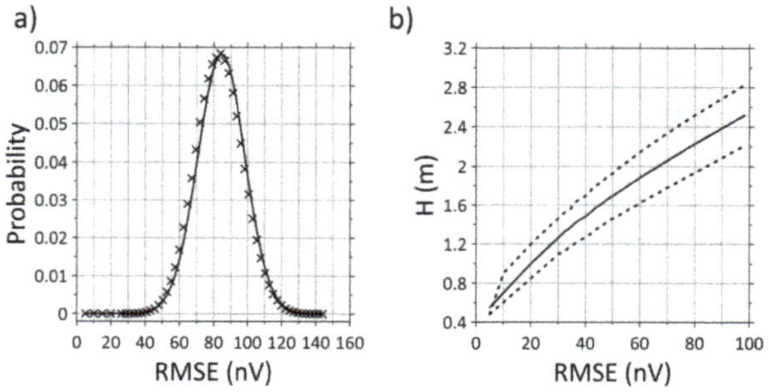


**Figure 6.23.** Smooth inversion of the noise contaminated synthetic data set with a 40-layer inverse model: (a) the probability density function versus RMSE computed directly (black crosses) and computed assuming the normal distribution using the mean and the standard deviation (solid black line); (b) the water column (black solid line) and the 95% confidence interval (black dashed lines) versus RMSE.

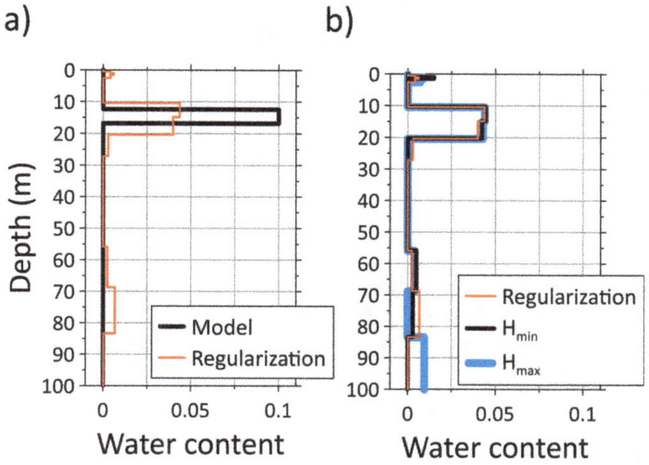


**Figure 6.24.** Inversion of the noise contaminated synthetic data set with a 17-layer inverse model: (a) the true model (black line) and the regularized inverse model (red line); (b) the regularized inverse model (red line) and the perturbed equivalent inverse models corresponding to the minimum (black line) and maximum (blue line) water columns solutions.

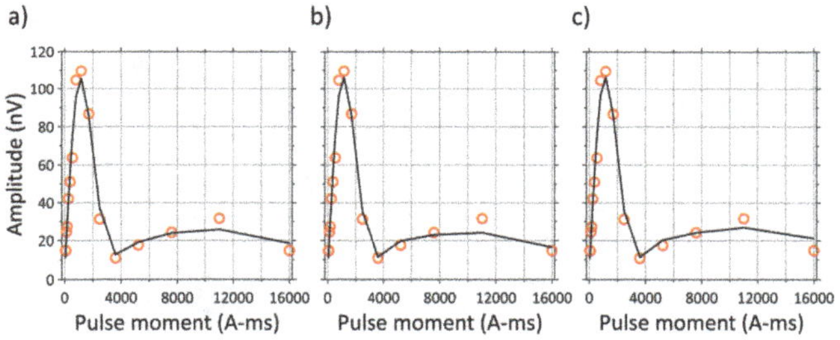


**Figure 6.25.** Inversion of the noise contaminated synthetic data set with a 17-layer inverse model. The signal amplitude (red circles) fitted by the theoretical signal (black line) corresponding to the equivalent inverse models: (a) regularized inversion; (b) the minimum water column inversion; (c) the maximum water column inversion.

The input data are well fitted by the inversion with the fitting error RMSE = 4.8nV (figure 6.25).

The probability function shows that the great majority of the perturbed inverse models do not fit the input data with the accuracy equal to the regularized inverse model (figure 6.26).

*6.2.1.3 Inversion with the minimum number of layers using a 12-layer model*

Inversion with the 12-layer inverse model shows the location of the shallow water model and underestimates the water content. Inversion suggests the existence of deep water. This deep water solution can be explained by the noise influence (figure 6.27).

The input data are well fitted by the inversion with the fitting error RMSE = 5.3nV (figure 6.28).

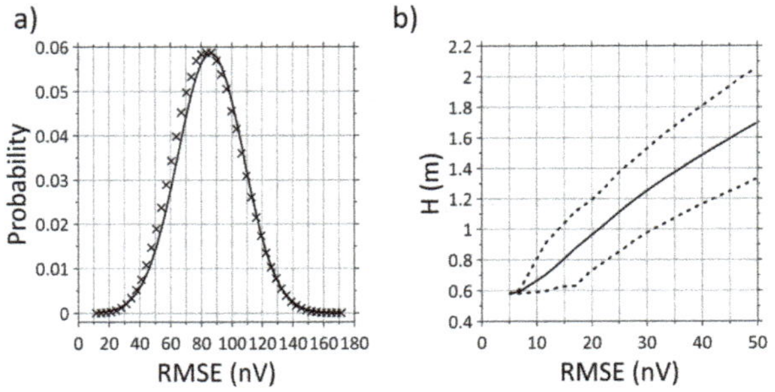


**Figure 6.26.** Inversion of the noise contaminated synthetic data set with a 17-layer inverse model: (a) the probability density function versus RMSE computed directly (black crosses) and computed assuming the normal distribution using the mean and the standard deviation (solid black line); (b) the water column (black solid line) and the 95% confidence interval (black dashed lines) versus RMSE.

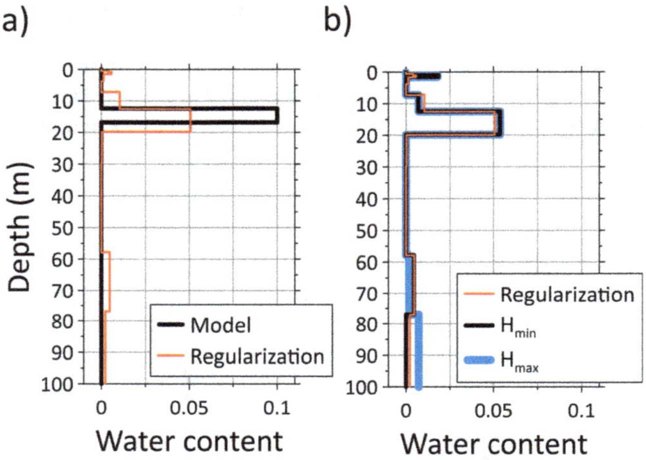


**Figure 6.27.** Inversion of the noise contaminated synthetic data set with a 12-layer inverse model: (a) the true model (black line) and the regularized inverse model (red line); (b) the regularized inverse model (red line) and the perturbed equivalent inverse models corresponding to the minimum (black line) and maximum (blue line) water columns solutions.

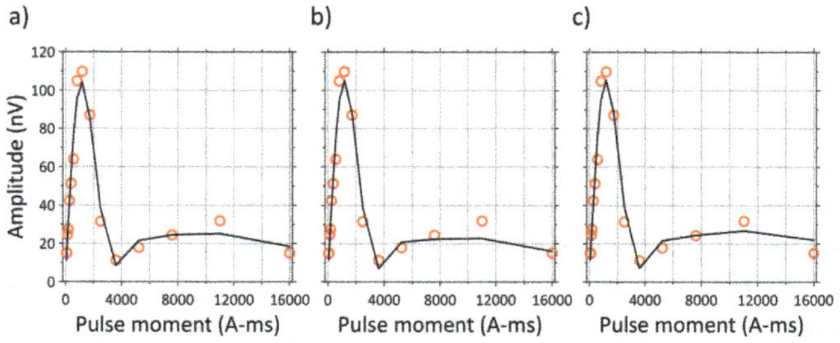


**Figure 6.28.** Inversion of the noise contaminated synthetic data set with a 12-layer inverse model. The signal amplitude (red circles) fitted by the theoretical signal (black line) corresponding to the equivalent inverse models: (a) regularized inversion; (b) the minimum water column inversion; (c) the maximum water column inversion.

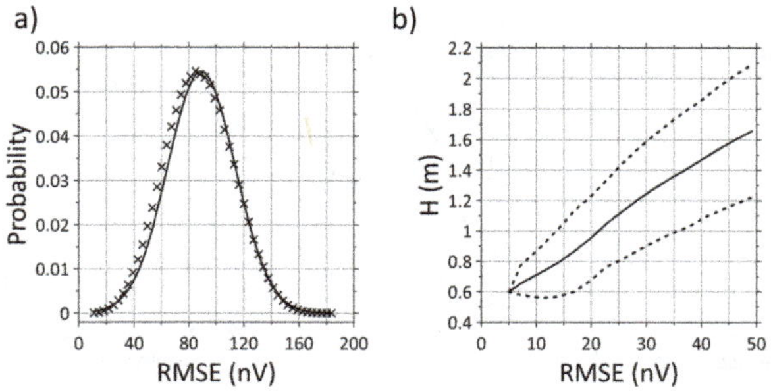


**Figure 6.29.** Inversion of the noise contaminated synthetic data set with a 12-layer inverse model: (a) the probability density function versus RMSE computed directly (black crosses) and computed assuming the normal distribution using the mean and the standard deviation (solid black line); (b) the water column (black solid line) and the 95% confidence interval (black dashed lines) versus RMSE.

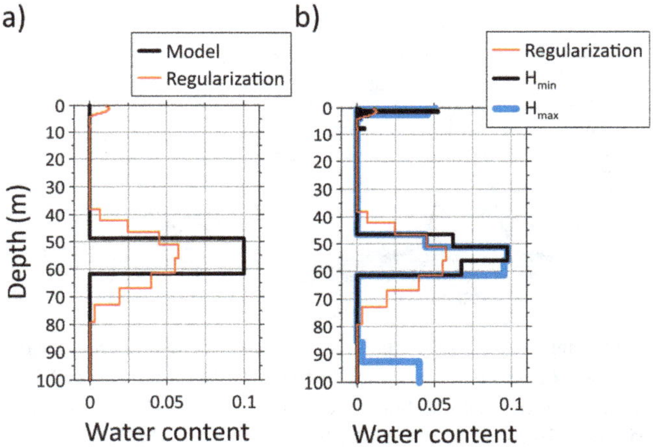


**Figure 6.30.** Smooth inversion of the noise contaminated synthetic data set with a 40-layer inverse model: (a) the true model (black line) and the regularized inverse model (red line); (b) the regularized inverse model (red line) and the perturbed equivalent inverse models corresponding to the minimum (black line) and maximum (blue line) water columns solutions.

The probability function shows that the great majority of the perturbed inverse models do not fit the input data with the accuracy equal to the regularized inverse model (figure 6.29).

### 6.2.2 Model: deep water

The noise standard deviation for the deep water model $\sigma = 14.55$nV which corresponds to 11% of the maximum amplitude of the MRS signal.

#### *6.2.2.1 Smooth inversion with a 40-layer model*

The regularized inverse model recovers location of the deep water model, but underestimates the water content. The $H_{\min}$ and $H_{\max}$ inverse models show better results, but produce the artifacts at the top and the bottom of the water content profile (figure 6.30).

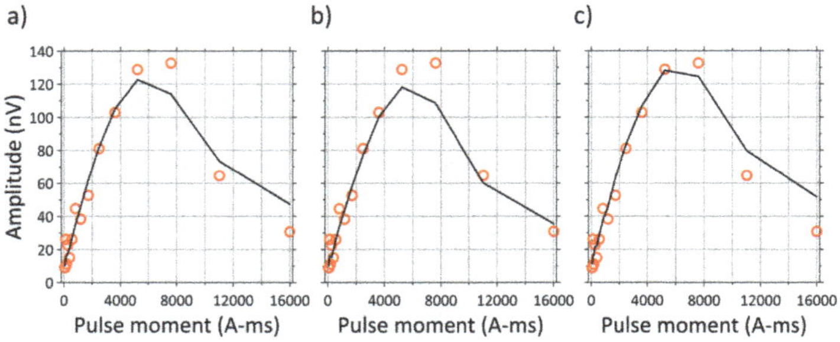


**Figure 6.31.** Smooth inversion of the noise contaminated synthetic data set with a 40-layer inverse model. The signal amplitude (red circles) fitted by the theoretical signal (black line) corresponding to the equivalent inverse models: (a) regularized inversion; (b) the minimum water column inversion; (c) the maximum water column inversion.

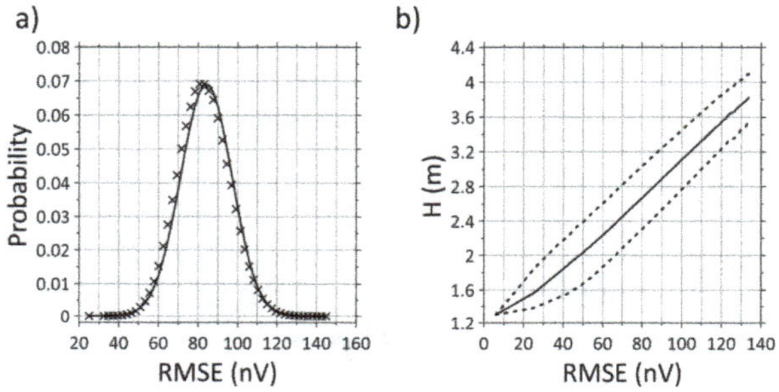


**Figure 6.32.** Smooth inversion of the noise contaminated synthetic data set with a 40-layer inverse model: (a) the probability density function versus RMSE computed directly (black crosses) and computed assuming the normal distribution using the mean and the standard deviation (solid black line); (b) the water column (black solid line) and the 95% confidence interval (black dashed lines) versus RMSE.

The input data are fitted by the inversion with the fitting error RMSE = 8.6nV (figure 6.31).

The probability function shows that the great majority of the perturbed inverse models do not fit the input data with the accuracy equal to the regularized inverse model (figure 6.32).

*6.2.2.2 Inversion with the maximum resolution using a 17-layer model*

The inverse model recovers the location of the deep water model, but underestimates the water content and produces the artifacts at the top and the bottom of the water content profile (figure 6.33).

The input data are fitted by the inversion with the fitting error RMSE = 9.2nV (figure 6.34).

The probability function shows that the great majority of the perturbed inverse models do not fit the input data with the accuracy equal to the regularized inverse model (figure 6.35).

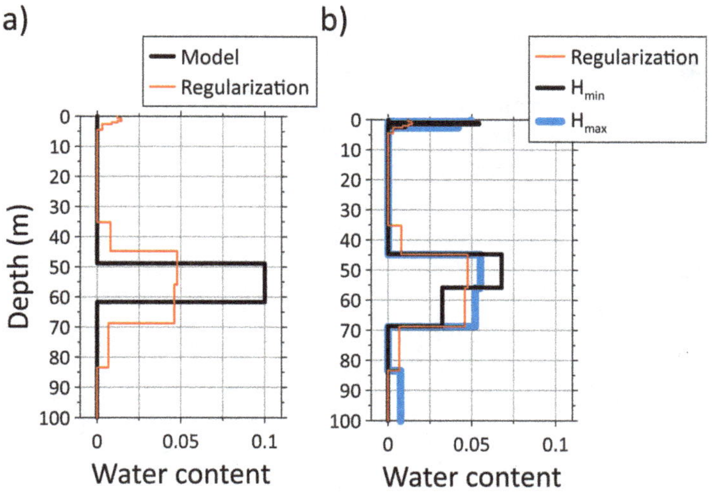


**Figure 6.33.** Inversion of the noise contaminated synthetic data set with a 17-layer inverse model: (a) the true model (black line) and the regularized inverse model (red line); (b) the regularized inverse model (red line) and the perturbed equivalent inverse models corresponding to the minimum (black line) and maximum (blue line) water columns solutions.

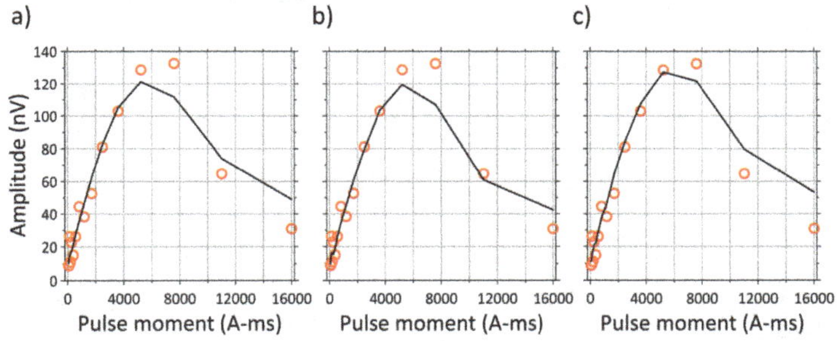


**Figure 6.34.** Inversion of the noise contaminated synthetic data set with a 17-layer inverse model. The signal amplitude (red circles) fitted by the theoretical signal (black line) corresponding to the equivalent inverse models: (a) regularized inversion; (b) the minimum water column inversion; (c) the maximum water column inversion.

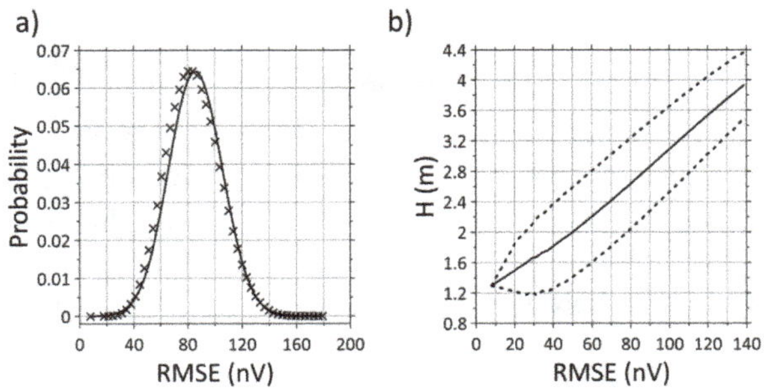


**Figure 6.35.** Inversion of the noise contaminated synthetic data set with a 17-layer inverse model: (a) the probability density function versus RMSE computed directly (black crosses) and computed assuming the normal distribution using the mean and the standard deviation (solid black line); (b) the water column (black solid line) and the 95% confidence interval (black dashed lines) versus RMSE.

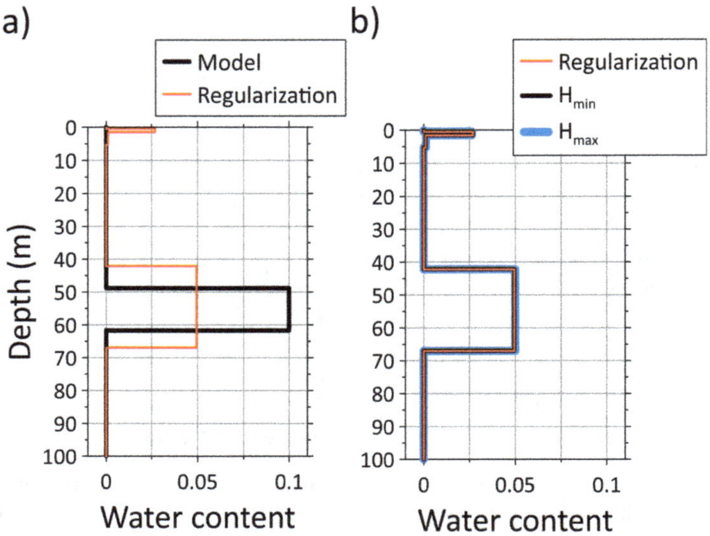


**Figure 6.36.** Inversion of the noise contaminated synthetic data set with an 8-layer inverse model: (a) the true model (black line) and the regularized inverse model (red line); (b) the regularized inverse model (red line) and the perturbed equivalent inverse models corresponding to the minimum (black line) and maximum (blue line) water columns solutions.

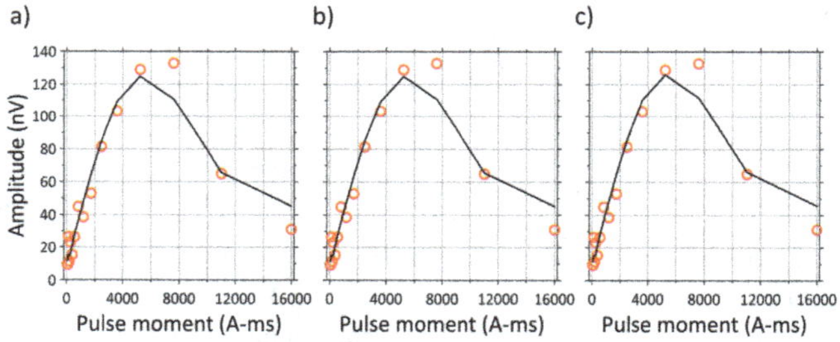


**Figure 6.37.** Inversion of the noise contaminated synthetic data set with an 8-layer inverse model. The signal amplitude (red circles) fitted by the theoretical signal (black line) corresponding to the equivalent inverse models: (a) regularized inversion; (b) the minimum water column inversion; (c) the maximum water column inversion.

### *6.2.2.3 Inversion with the minimum number of layers using an 8-layer model*

The inverse model recovers the location of the deep water model, but underestimates the water content. The 8-layer inversion is more stable than the multi-layer inversions and produces the artifacts only at the top of the water content profile (figure 6.36).

The input data are fitted by the inversion with the fitting error RMSE = 9.0nV (figure 6.37).

The probability function shows that the great majority of the perturbed inverse models do not fit the input data with the accuracy equal to the regularized inverse model (figure 6.38).

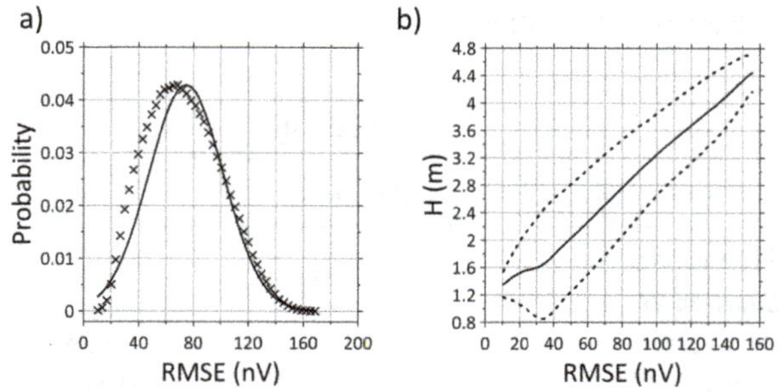


**Figure 6.38.** Inversion of the noise contaminated synthetic data set with an 8-layer inverse model: (a) the probability density function versus RMSE computed directly (black crosses) and computed assuming the normal distribution using the mean and the standard deviation (solid black line); (b) the water column (black solid line) and the 95% confidence interval (black dashed lines) versus RMSE.

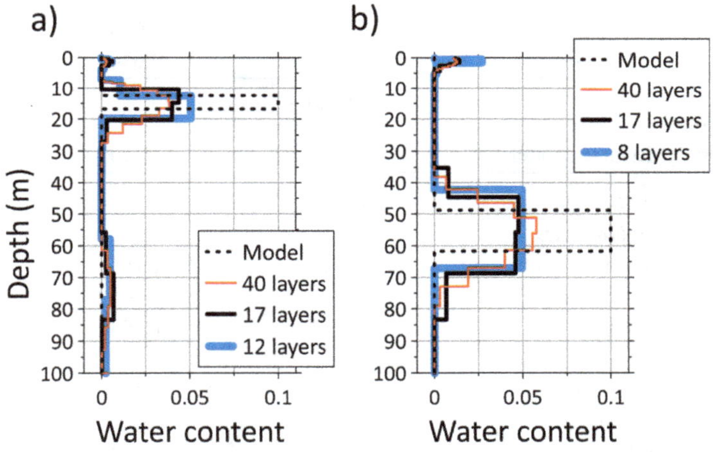


**Figure 6.39.** Inversion of the noise contaminated synthetic data sets: (a) the shallow water model of the water content (black dashed line) and three inverse models corresponding to the regularized solution using 40, 17 and 12 model layers; (b) the deep water model (black dashed line) and three inverse models corresponding to the regularized solution using 40, 17 and 8 model layers.

### 6.2.3 Summary

Inversion of the noise contaminated synthetic data sets corresponding to the shallow and the deep water models and carried out with different number of model layers shows that the models are recovered with equal accuracy. Figure 6.39 shows that the inverse models underestimate double the water content and overestimate the thickness of the model.

The results of the noise contaminated data inversion are summarized in table 6.2.

Figure 6.40 shows the relative error of the water column estimate versus the number of the inverse model in table 6.2. The error for the inverse models 1–9 is

**Table 6.2.** The water column ($H$), the RMSE and the relative error of the water column estimates for the noiseless data sets.

| | Model: shallow water | | | |
|---|---|---|---|---|
| Number | Inverse model | RMSE (nV) | $H$ (m) | Error $H$ (%) |
| | True model | | 0.442 | |
| | ***Solution $H_{min}$*** | | | |
| 1 | 40 layers | 5.69 | 0.435 | −1.5 |
| 2 | 17 layers | 4.84 | 0.544 | 23.1 |
| 3 | 12 layers | 5.32 | 0.529 | 19.7 |
| | ***Regularization*** | | | |
| 4 | 40 layers | 5.70 | 0.563 | 27.5 |
| 5 | 17 layers | 4.86 | 0.574 | 30.0 |
| 6 | 12 layers | 5.32 | 0.580 | 31.2 |
| | ***Solution $H_{max}$*** | | | |
| 7 | 40 layers | 5.61 | 0.834 | 88.6 |
| 8 | 17 layers | 4.85 | 0.648 | 46.7 |
| 9 | 12 layers | 5.31 | 0.644 | 45.7 |
| | Model: deep water | | | |
| | Inverse model | RMSE (nV) | $H$ (m) | Error $H$ (%) |
| | True model | | 1.285 | |
| | ***Solution $H_{min}$*** | | | |
| 10 | 40 layers | 8.53 | 1.164 | −9.4 |
| 11 | 17 layers | 9.12 | 1.204 | −6.3 |
| 12 | 8 layers | 9.00 | 1.258 | −2.1 |
| | ***Regularization*** | | | |
| 13 | 40 layers | 8.64 | 1.310 | 1.9 |
| 14 | 17 layers | 9.33 | 1.328 | 3.4 |
| 15 | 8 layers | 9.00 | 1.258 | −2.1 |
| | ***Solution $H_{max}$*** | | | |
| 16 | 40 layers | 8.61 | 1.550 | 20.6 |
| 17 | 17 layers | 9.19 | 1.452 | 13.0 |
| 18 | 8 layers | 9.00 | 1.270 | −1.2 |

larger than that for the inverse models 10–18, thus showing that the water column is better estimated for the deep water model composed of a thick water saturated layer.

The $H_{min}$ and $H_{max}$ inverse models are useful for investigating uncertainty in the inversion results. Knowledge of the uncertainty may allow better interpretation of field measurements, which is important when borehole data or other information about the investigated subsurface is available. The estimate of the uncertainty also allows to establish how much the modeling parameters can be varied when using MRS results for hydrodynamic modeling. The regularized inverse model is a typical result of the MRS study.

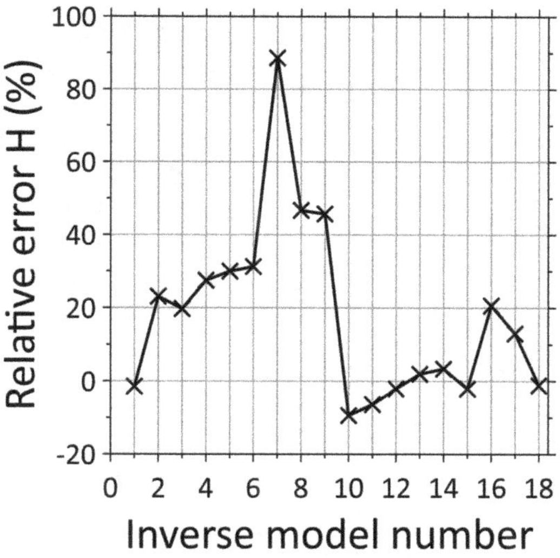


**Figure 6.40.** The relative error of the water column estimates for the noise contaminated data sets versus the inverse model number.

**Table 6.3.** The relative error of the water content (*w*) and the water column (*H*) estimates for the regularized inverse models.

| Number | Inverse model | Error *w* (%) | Error *H*(%) | Data set |
|---|---|---|---|---|
| 1 | 40 layers | −23.0 | 10.2 | Shallow, noiseless |
| 2 | 17 layers | −45.3 | 11.6 | Shallow, noiseless |
| 3 | 12 layers | −48.3 | 7.5 | Shallow, noiseless |
| 4 | 40 layers | −5.5 | 2.5 | Deep, noiseless |
| 5 | 17 layers | −40.0 | 2.5 | Deep, noiseless |
| 6 | 8 layers | −50.0 | 4.0 | Deep, noiseless |
| 7 | 40 layers | −62.0 | 27.5 | Shallow, noise contaminated |
| 8 | 17 layers | −56.0 | 30.0 | Shallow, noise contaminated |
| 9 | 12 layers | −49.0 | 31.2 | Shallow, noise contaminated |
| 10 | 40 layers | −42.0 | 1.9 | Deep, noise contaminated |
| 11 | 17 layers | −53.0 | 3.4 | Deep, noise contaminated |
| 12 | 8 layers | −51.0 | −2.1 | Deep, noise contaminated |

It has been shown that the most stable parameter provided by MRS inversion is the water column. Table 6.3 summarizes the relative error of the water content and the water column estimates considering inversion of the synthetic data discussed above.

Figure 6.41 visualizes data presented in table 6.3 and shows that for our examples, the errors of the water column estimate are at least twice smaller than that for the water content.

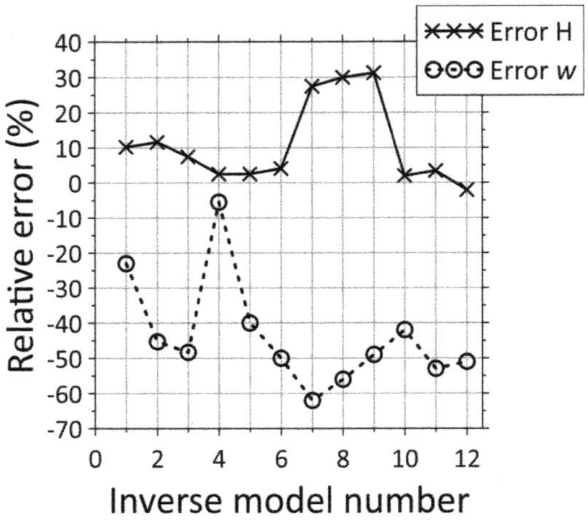


**Figure 6.41.** The relative error of the water content ($w$) and the water column ($H$) estimates for the regularized inverse models versus the inverse model number.

## 6.3 Field example

Now, we present an example of the MRS field study carried out in the test site near St-Cyr-en-Val village in France. The MRS test area is part of the Beauce aquifer (9000 km$^2$) located south-west of Paris near Orleans city (figure 6.42).

The mean annual recharge is 110 mm. Developed agriculture with irrigation in summer is widespread in the area. Soil thickness ranges from 0.3 to 1.5 m. The water column in these soils varies between 50 and 180 mm. The limestone underlying the soil layer is heterogeneous with the variable fracturing and irregular karst development. Estimations of the specific yield of the limestone show 3 to 13%. In the test area of St-Cyr-en-Val, three aquifers were determined during a previous geological study. A stratum of mixed gravel, sand and clay, approximately 20–25 m thick, make the upper aquifer. A water-saturated karst limestone separated by a sandstone layer at a depth of 50–60 m makes the two others. The total thickness of the karst limestone aquifers is about 40–50 m, and the average porosity is approximately 10%. The static water level measured in borehole N 268 situated approximately 600 m from the MRS test site showed 4 m. The lithological log, the resistivity distribution measured with the GEONICS PROTEM TDEM instrument and measured MRS amplitudes versus pulse moment are presented in figure 6.43.

Presented MRS sounding was performed using the NUMIS instrument with a 100 × 100 m$^2$ square loop and the maximum pulse moment of 9100 A-ms. The Larmor frequency was 2014 Hz. The average electromagnetic noise after stacking was measured at 52 nV, which makes about 10% of the maximum signal amplitude. The local Earth's magnetic field was 47 214 nT with the inclination of 63°N, and the resistivity of the subsurface was provided by the TDEM measurements.

Inversion of this data set is carried out using the SAMOVAR software. First, we estimate the maximum depth of water detection by comparing the amplitude of the MRS signal generated by a 1-m-thick layer of bulk water with the noise level. In MRS, the noise amplitude is measured before the pulse and noise may be different in

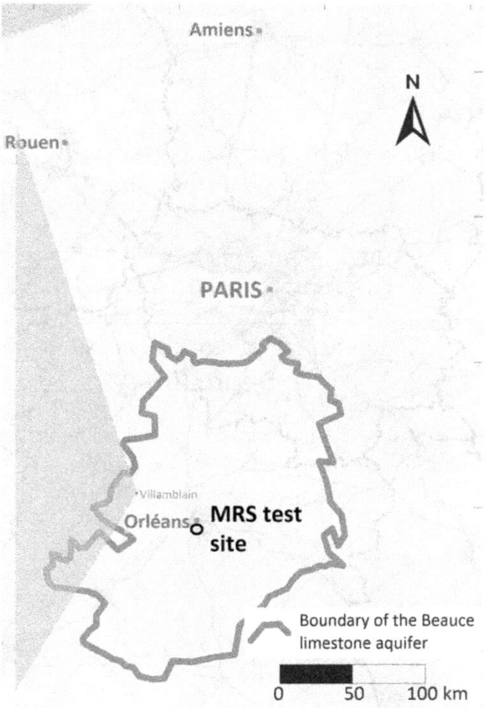


**Figure 6.42.** Location of the test area of St-Cyr-en-Val near Orleans, France.

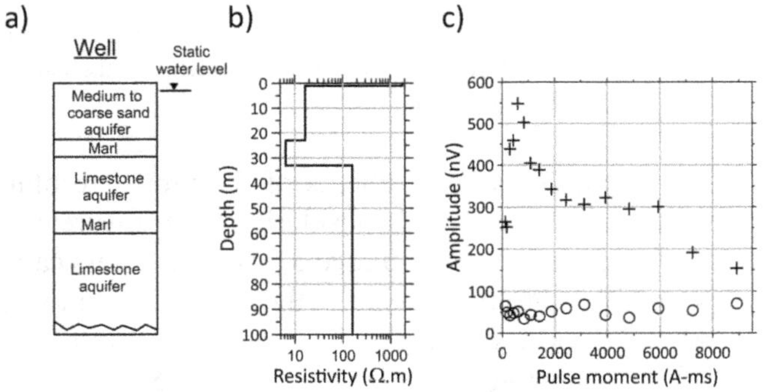


**Figure 6.43.** The input data in the test site of St-Cyr-en-Val: (a) the lithological log of borehole N 268; (b) the resistivity distribution; (c) the amplitude of MRS signal versus pulse moment.

comparison with that contaminating the MRS signal measured after the pulse. The noise phase and the signal phase are not correlated and the noise estimate for inversion is computed by multiplying the measured noise amplitude by 0.7. Hence, noise is known only approximately. Figure 6.44 shows the amplitude generated by a 1-m-thick layer of bulk water versus depth compared with the noise estimate $\delta = 52 \times 0.7 \approx 36$ nV. At the depth of 80 m the noise and the signal are equal and we consider this depth as the maximum depth of investigation.

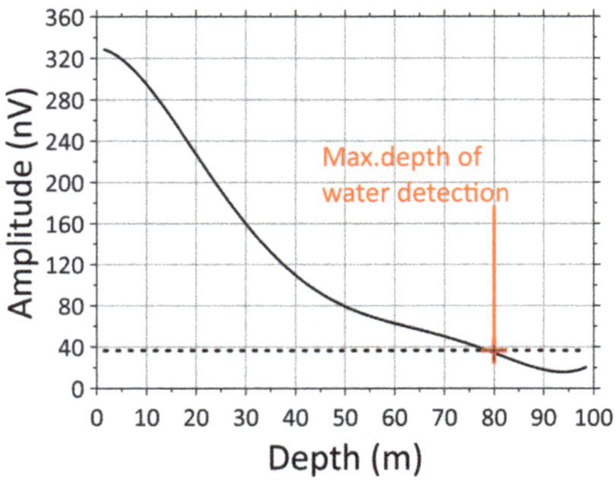


**Figure 6.44.** The amplitude generated by a 1-m-thick layer of bulk water versus depth under St-Cyr-en-Val measuring conditions. The dashed line shows the noise level δ = 36 nV.

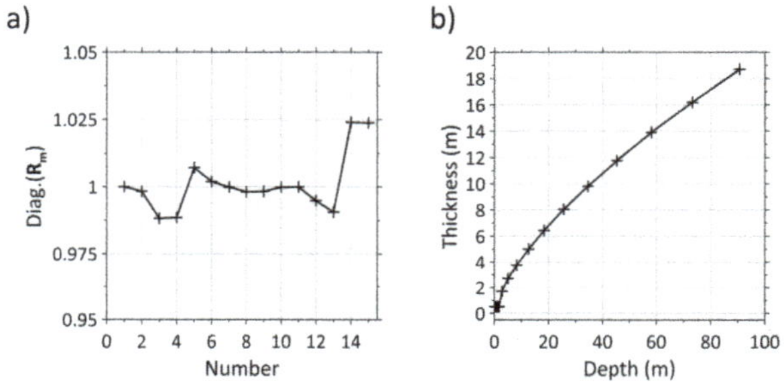


**Figure 6.45.** The maximum theoretical resolution under St-Cyr-en-Val measuring conditions computed considering discretization with 15 model layers: (a) the diagonal elements of the model resolution matrix; (b) the model layer thickness versus the layer depth.

Figure 6.45(a) shows the diagonal elements of the model resolution matrix $\mathbf{R_m}$ corresponding to the matrix **A** computed considering the measurement conditions in the test site and the discretization with 15 layers. Over 15 layers cannot be resolved and hence 15 layers correspond to the maximum theoretical resolution of the matrix **A**. Figure 6.45(b) shows the thickness of these 15 layers versus depth.

We make the first inversion using the inverse model with the maximum theoretical resolution and applying regularization. The uncertainty estimate is done assuming homogeneous Gaussian noise. Figure 6.46 shows the 95% confidence interval for the regularized solution computed considering the inverse model with 15 layers. The uncertainty for this model is minimal for layers between 20 and 30 m and increases for both shallow and deep layers. However, figure 6.43(c) shows that the noise amplitude measured for each value of the pulse moment was varying between 35 and 70 nV, which does not correspond to the homogeneity assumption. This suggests to use the average uncertainty estimate equal to $\Delta w_{n,j} \approx 0.1$ for any $n$ and $j$ in equation (5.29).

We estimate the uncertainty in the regularized model by applying random perturbations. The amplitude of the perturbation $\Delta w_{n,j}$ can be set either equal to

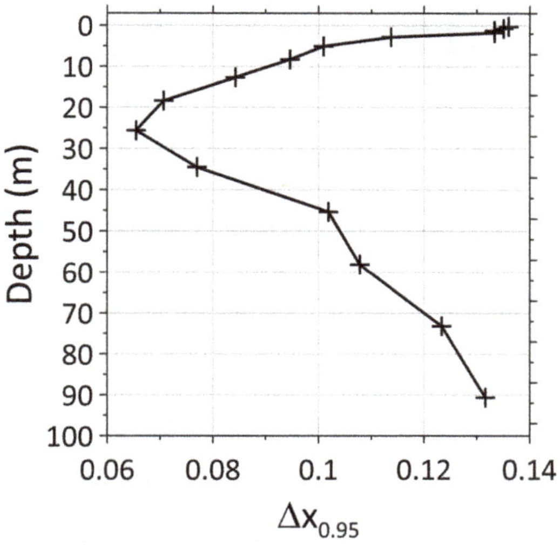


**Figure 6.46.** The 95% confidence interval for the regularized solution computed considering the St-Cyr-en-Val measuring conditions and the inverse model with 15 layers.

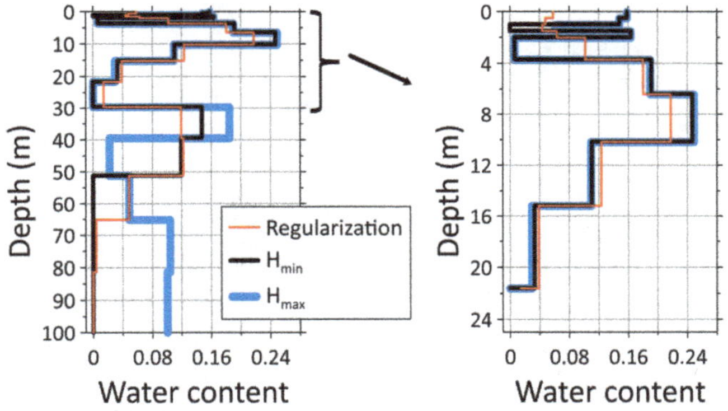


**Figure 6.47.** Inversion of the field data with a 15-layer inverse model: the regularized inverse model (red line) and the perturbed equivalent inverse models corresponding to the minimum (black line) and maximum (blue line) water columns solutions. The right graph shows a zoom of the shallow part of the water content profile.

the SVD estimate of the 95% confidence interval for each model layer or we can use an arbitrary value equal for all layers. For this data set, we use the average value $\Delta w_{n,j} = 0.1$. Figure 6.47 shows the regularized and perturbed inverse models.

The shallow aquifer between 4 and 15 m is shown by all three inverse models. An impermeable layer separates this aquifer and the second aquifer between 30 and 50 m. The deep aquifer below 65 m is suggested by the $H_{\max}$ solution, but not shown by two other equivalent inverse models. Hence, MRS does not exclude a deep water-saturated formation at the limit of the water detection capacity of the method, but it would be erroneous to insist that MRS points to this aquifer. Figure 6.48 shows the measured amplitudes (red circles) fitted by the equivalent inverse models with the RMSE = 27.7nV.

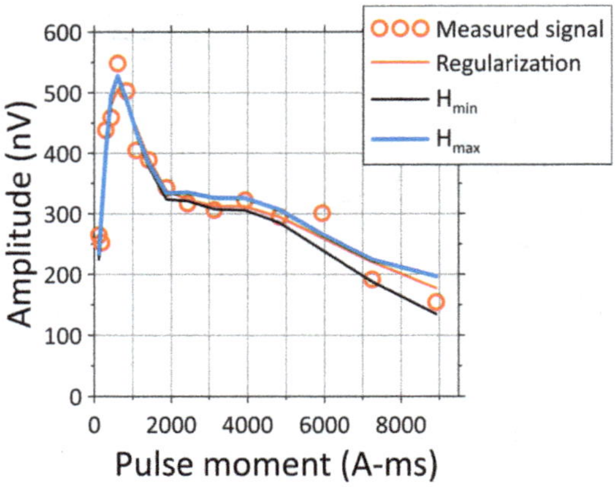


**Figure 6.48.** Inversion of the field data with a 15-layer inverse model. The measured signal amplitude (red circles) fitted by the theoretical signals corresponding to the equivalent inverse models: regularized inversion (red line), the minimum water column inversion (black line) and the maximum water column inversion (blue line).

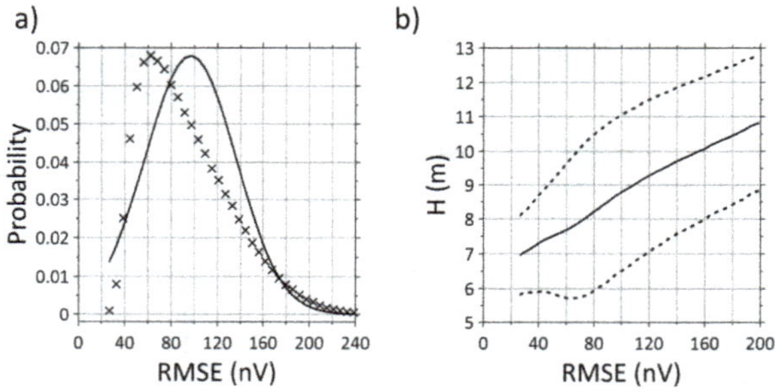


**Figure 6.49.** Inversion of the field data with a 15-layer inverse model: (a) the probability density function versus RMSE computed directly (black crosses) and computed assuming the normal distribution using the mean and the standard deviation (solid black line); (b) the water column (black solid line) and the 95% confidence interval (black dashed lines) versus RMSE.

The probability density function shown in figure 6.49 is different from the normal distribution typically shown with the synthetic data. This suggests that the real electromagnetic noise was not normally distributed.

The smooth inversion of the field data using a 40-layer inverse model (figures 6.50–6.52) and using a 9-layer inverse model corresponding to the minimum number of model layers (figures 6.53–6.55) show the MRS water content logs consistent with that given by the 15-layer inverse model. The field data are fitted well with the RMSE = 24.3nV (40-layer inverse model) and RMSE = 28nV (9-layer inverse model), which is less than the average noise amplitude estimated at 36 nV. Hence, all these inverse models represent the equivalent solutions.

The multi-layer inverse models show two aquifers. The first aquifers between approximately 5 and 15 m and the second aquifer between 30 and 50 m. A waterless interval between 15 and 30 m separates these aquifers. The deep aquifer is shown only by the maximum water column solutions and should be considered as possibly

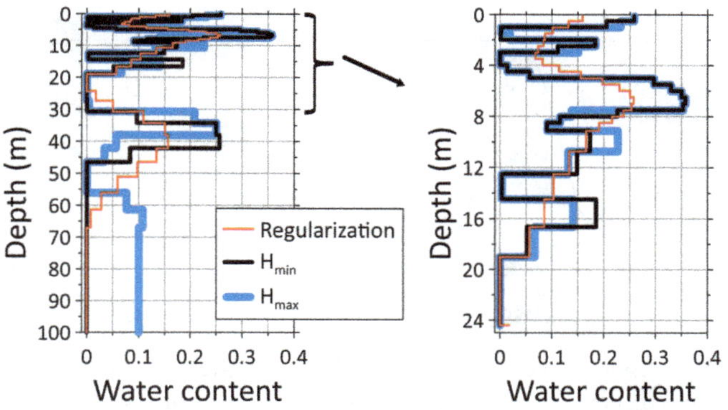


**Figure 6.50.** Smooth inversion of the field data with a 40-layer inverse model: the regularized inverse model (red line) and the perturbed equivalent inverse models corresponding to the minimum (black line) and maximum (blue line) water columns solutions. The right graph shows a zoom of the shallow part of the water content profile.

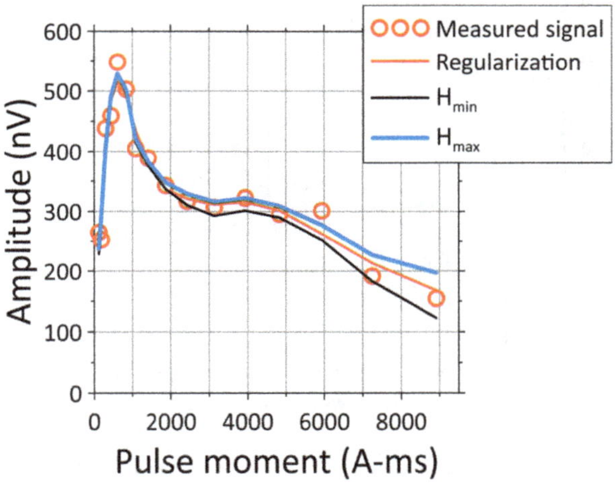


**Figure 6.51.** Smooth inversion of the field data with a 40-layer inverse model. The signal amplitude (red circles) fitted by the theoretical signal (black line) corresponding to the equivalent inverse models: (a) regularized inversion; (b) the minimum water column inversion; (c) the maximum water column inversion.

observed. These results suggest trying the blocky inversion assuming two layers. In this inversion scheme, the boundaries and the water content for each layer are optimized by inversion. Figure 6.56 shows the 2-layer inverse model (3.75–15.5 m and 30–49.75 m) that fits experimental data with the RMSE = 32.6nV (figure 6.57). The fitting error is less than the noise estimate and this equivalent inverse model confirms that the deep aquifer is not seen by MRS.

Now, we should also verify whether the waterless interval between 15 and 30 m is reliably detected by MRS or an equivalent inverse model without this interval also

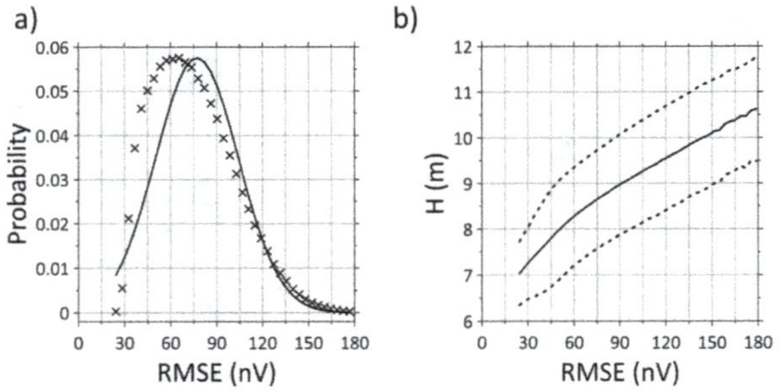


**Figure 6.52.** Smooth inversion of the field data with a 40-layer inverse model: (a) the probability density function versus RMSE computed directly (black crosses) and computed assuming the normal distribution using the mean and the standard deviation (solid black line); (b) the water column (black solid line) and the 95% confidence interval (black dashed lines) versus RMSE.

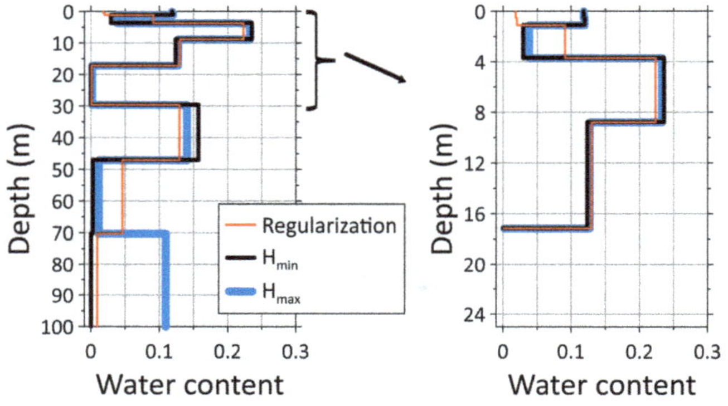


**Figure 6.53.** Inversion of the field data with a 9-layer inverse model: the regularized inverse model (red line) and the perturbed equivalent inverse models corresponding to the minimum (black line) and maximum (blue line) water columns solutions. The right graph shows a zoom of the shallow part of the water content profile.

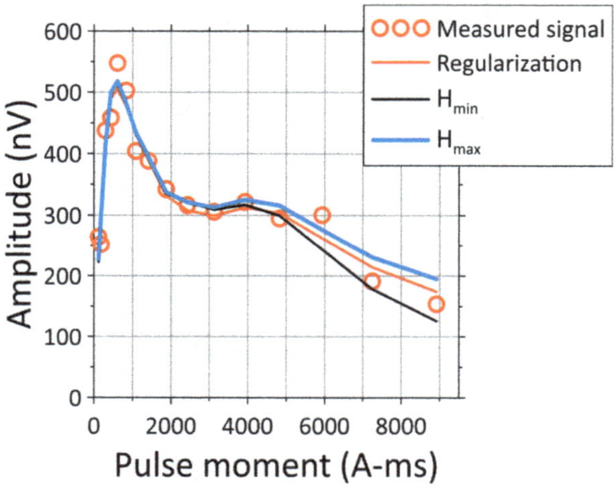


**Figure 6.54.** Inversion of the field data with a 9-layer inverse model. The signal amplitude (red circles) fitted by the theoretical signal (black line) corresponding to the equivalent inverse models: (a) regularized inversion; (b) the minimum water column inversion; (c) the maximum water column inversion.

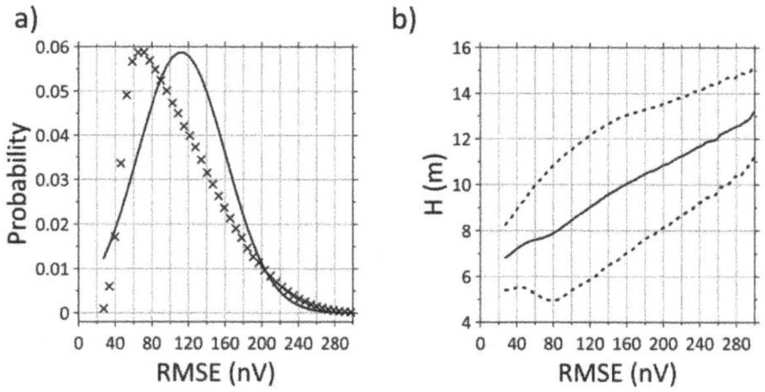


**Figure 6.55.** Inversion of the field data with a 9-layer inverse model: (a) the probability density function versus RMSE computed directly (black crosses) and computed assuming the normal distribution using the mean and the standard deviation (solid black line); (b) the water column (black solid line) and the 95% confidence interval (black dashed lines) versus RMSE.

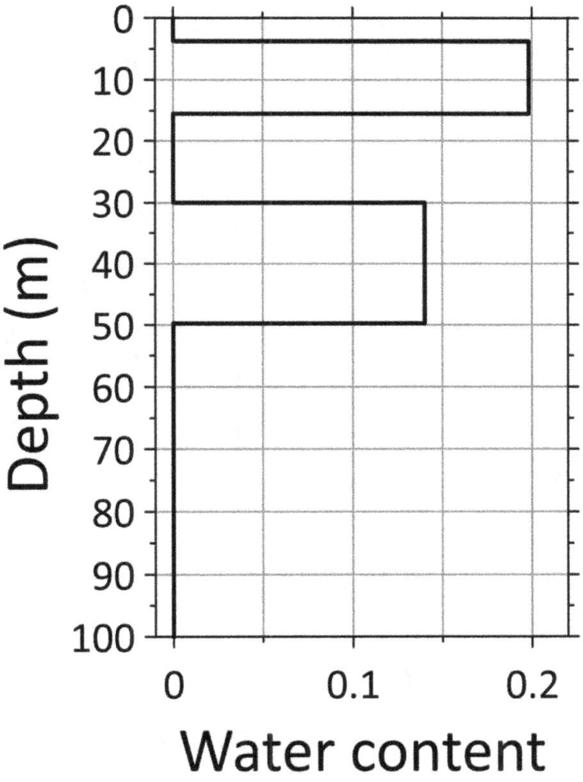


**Figure 6.56.** Blocky inversion of the field data with a 2-layer inverse model.

exists. For that, we use a 1-layer inverse model and the blocky inversion routine. Figure 6.58 shows the model that fits the best field data.

The 1-layer inverse model allow fitting experimental data with the RMSE = 69.6nV (figure 6.59), which is larger than the noise amplitude (36 nV)

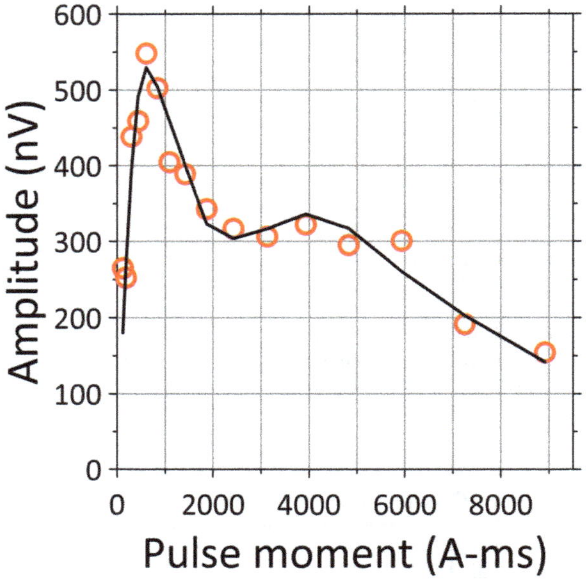


**Figure 6.57.** Blocky inversion of the field data with a 2-layer inverse model. The signal amplitude (red circles) fitted by the theoretical signal (black line).

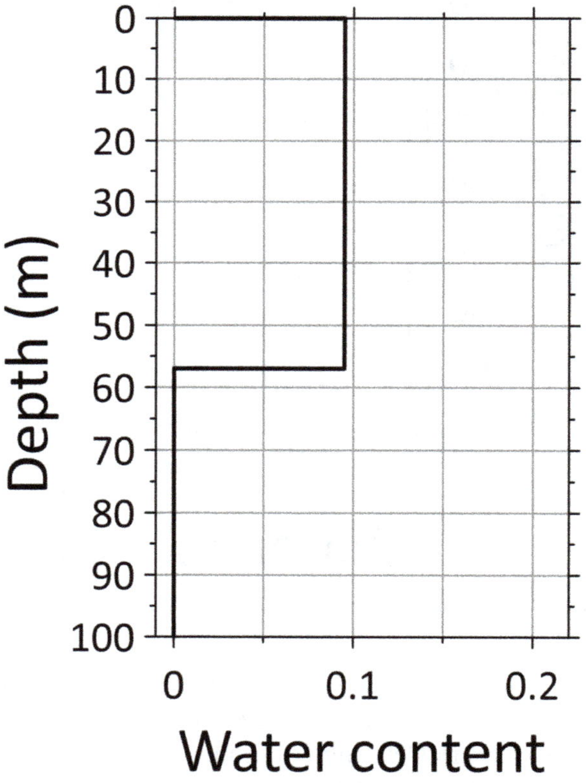


**Figure 6.58.** Blocky inversion of the field data with a 1-layer inverse model.

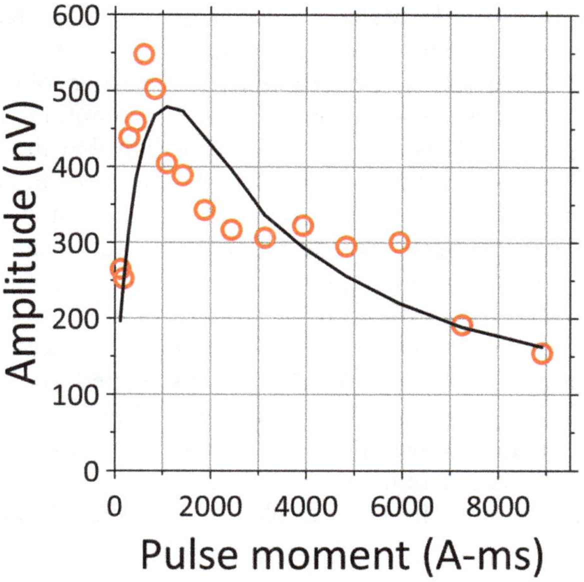


**Figure 6.59.** Blocky inversion of the field data with a 1-layer inverse model. The signal amplitude (red circles) fitted by the theoretical signal (black line).

**Table 6.4.** Inversion of the field measurements in the test site of St-Cyr-en-Val. The water column estimate $H$ is computed using the inverse models with different number of model layers and the deviation of the water column estimate $\Delta H$ is computed relative to the mean value of the water column $H_{mean}$.

| | | First aquifer | | Second aquifer | |
|---|---|---|---|---|---|
| *Number* | *Inverse model* | $H$ (m) | $\Delta H$ (%) | $H$ (m) | $\Delta H$ (%) |
| 1 | 40 layers $H_{min}$ | 2.53 | 2.11 | 2.71 | −8.71 |
| 2 | 15 layers $H_{min}$ | 2.46 | −0.92 | 2.85 | −3.95 |
| 3 | 9 layers $H_{min}$ | 2.44 | −1.50 | 2.80 | −5.76 |
| 4 | 40 layers regularization | 2.47 | −0.32 | 3.33 | 12.18 |
| 5 | 15 layers regularization | 2.57 | 3.52 | 3.34 | 12.66 |
| 6 | 9 layers regularization | 2.49 | 0.57 | 3.32 | 11.95 |
| 7 | 2 layers block inversion | 2.33 | −6.01 | 2.77 | −6.72 |
| 8 | 40 layers $H_{max}$ | 2.53 | 2.11 | 3.09 | 4.09 |
| 9 | 15 layers $H_{max}$ | 2.46 | −0.62 | 2.75 | −7.36 |
| 10 | 9 layers $H_{max}$ | 2.50 | 1.05 | 2.72 | −8.39 |
| | $H_{mean}$ | 2.48 | | 2.97 | |

and hence, this inverse model is unacceptable as the inversion results. We may conclude that the MRS inversion reliably shows two aquifers separated by the waterless interval (figure 6.58).

Table 6.4 shows the water column computed for the two detected aquifers. These aquifers are hydraulically separated and we compute the water column in each of them. The mean value of the water column $H_{mean}$ is computed considering all presented inverse models. Note that we compute $H_{max}$ and $H_{min}$ inverse models considering the total investigated depth (0–100 m). The water column in each aquifer individually may not correspond to the max–min assumption. For example, the water column in model N°1 is shown to be larger than that in model N°4. When using field data, the subsurface is known only approximately and the true values of the water column are unknown. For this reason, we cannot estimate the accuracy of the water column measurements. The deviation of the water column relative to the average value shows stability of the inversion and not the accuracy of the water column measurement.

Figure 6.60 shows the water column estimates presented in table 6.4. Figure 6.60(a) shows the water column estimates plotted versus the number of inverse model, and figure 6.60(b) shows the deviation of the water column relative to the mean value computed individually for the first and for the second aquifers.

The last step of the field data interpretation comprises a comparison of the MRS results with the ground truth. In this example, the ground truth is represented by the

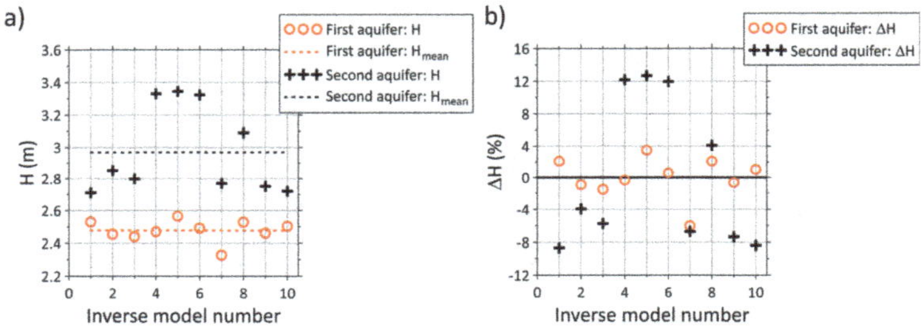


**Figure 6.60.** Inversion of the field measurements in the test site of St-Cyr-en-Val: (a) the water column estimates $H$ versus the number of inverse model; (b) the deviation of the water column estimate $\Delta H$ computed relative to the mean value of the water column.

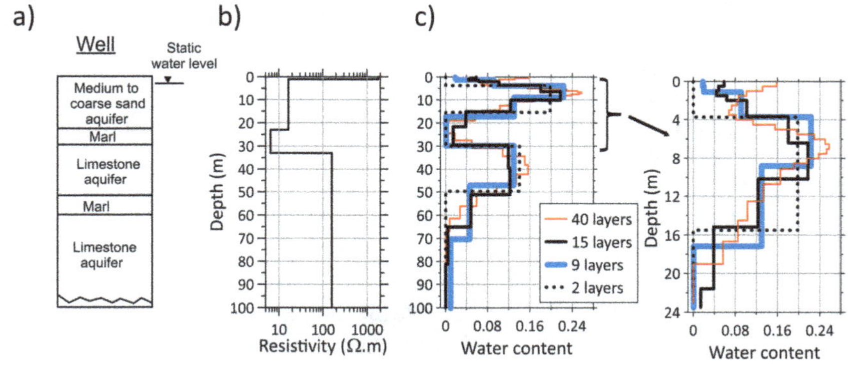


**Figure 6.61.** MRS study in the test site of St-Cyr-en-Val: (a) the lithological log of borehole N 268; (b) the resistivity distribution provided by the TDEM sounding; (c) the water content versus depth shown by the MRS inversion using the inverse models with different number of model layers.

lithological log and the resistivity log (figure 6.61(a) and figure 6.61(b)). The water column cannot be verified because of the impossibility of the water column measurements carried out with other methods at the scale of the MRS loop.

Figure 6.61(c) shows the water content logs given by different inverse models. The inverse models are consistent inbetween. MRS detects two aquifers separated by a waterless layer between approximately 15 and 30 m. This layer corresponds to a marl formation shown by the borehole log. In the first aquifer composed of medium to coarse sand, MRS shows the water content about 0.22, which is close to typical porosity for sand. The water content of 0.12 is a reasonable value for limestone. Below 50 m, MRS does not point to an aquifer shown by borehole. Some equivalent inverse models show this aquifer and hence, MRS does not exclude deep water. This deep aquifer is a well-known regional aquifer but MRS has not enough sensitivity for detecting this aquifer under test site measuring conditions.

**IOP** Publishing

Surface NMR for Hydrogeology
A user's guide
**Anatoly Legchenko**

# Chapter 7

# MRS and hydrogeology

## 7.1 MRS as a tool for hydrogeology

Magnetic resonance sounding (MRS) is distinguished from other geophysical tools by its selective sensitivity to groundwater. The magnetic resonance signal is generated by water molecules and the detection of groundwater with MRS corresponds to direct measurements. For example, in contrast to electrical resistivity measurements, inversion of MRS measurements reveals variations in water saturation that can be easier integrated into hydrogeological tools than the resistivity variations. Geometry of investigated aquifers and the quantity of water stored in the aquifer system is given by the MRS inversion for water content. It is discussed in the previous chapters. Here, we present in more details an estimation of the hydraulic conductivity, which is an important aquifer parameter.

The saturated hydraulic conductivity can be estimated using results of the MRS inversion calibrated with pumping tests in a borehole. The water content and an estimate of the hydraulic conductivity provided by MRS allow predicting the yield of water supply wells. MRS allows mapping of the hydraulic parameters of the subsurface at the catchment scale. Such a map may help to reduce the number of boreholes required for hydrogeological modeling and to interpolate aquifer hydraulic properties between boreholes. In rock with a low magnetic susceptibility (limestone, chalk), time-lapse MRS measurements allow a quantification of water stored in the unsaturated zone thus contributing to a quantification of the aquifer recharge. Note that the type of water containing material influence MRS signal, thus rendering the accuracy of the water content measurements in soils and rocks site dependent. The MRS water content is given by the inversion of field measurements, but estimation of the hydraulic conductivity with MRS is based on the empirical correlations between the relaxation times and the mean pore size of water-saturated rocks. Typical values of the MRS estimated water content and the relaxation times measured with MRS in different rocks are presented in table 7.1.

doi:10.1088/978-0-7503-3155-5ch7

**Table 7.1.** Typical values of measured with MRS water content and relaxation time $T_1$.

| Rock material | MRS water content | Relaxation time $T_1$ (ms) |
|---|---|---|
| Basalt, sand with >2% magnetite | 0 | |
| Clay | <0.03 | <100 |
| Dry sand, unsaturated weathered granite, solid granite | <0.01 | <100 |
| Unsaturated zone: chalk, limestone | 0.02–0.08 | <100 |
| Fissured granite, sandstone, non-fractured limestone | 0.02–0.08 | <200 |
| Highly fractured granite, sandstone, limestone | 0.02–0.08 | 350–600 |
| Karst | 0.02–0.03 | >400 |
| Compact chalk | 0.15–0.2 | <150 |
| Highly fractured chalk | 0.15–0.2 | 350–600 |
| Fine sand | 0.1–0.25 | <200 |
| Medium sand | 0.1–0.35 | 200–250 |
| Coarse sand | 0.2–0.4 | 250–500 |
| Gravel | 0.2–0.4 | >500 |

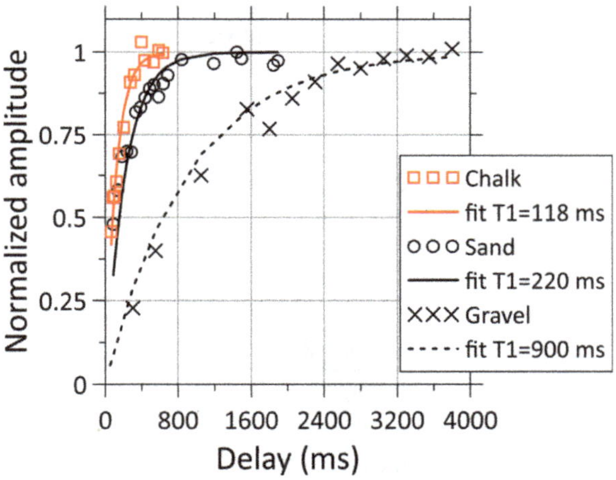


**Figure 7.1.** Normalized amplitude of the FID2 signal ($v_{02}$) measured after the second pulse versus delay between the pulses. Amplitude of the FID1 signal ($v_{01}$) is independent of the delay ($v_{01}$ = 1).

All three relaxation times $T_1$, $T_2$ and $T_2^*$ can be used for estimating the hydraulic conductivity. However, $T_2$ can be measured only when the Earth's magnetic field is not homogeneous. It happens in the presence of magnetic rocks. $T_2^*$ is very sensitive to variation of the magnetic susceptibility of water-bearing material. It can be used only in rocks with a low magnetic susceptibility (chalk, limestone). When rocks are slightly magnetic but the FID signal can be measured, $T_1$ is the most reliable estimator. Figure 7.1 shows an example of field measurements of the FID2 amplitude versus the delay between two pulses. At two sites in the same geological formation, the subsurface is composed of clay, chalk, sand and gravel. At the first site, MRS shows $T_1 = 118$ ms in compact chalk and $T_1 = 220$ ms in fine sand. At the second site, $T_1 = 900$ ms is measured in gravel. The yield of the boreholes drilled at each site shows 3 $m^3\ h^{-1}$ and 120 $m^3\ h^{-1}$ respectively.

The hydraulic conductivity can be estimated using equation (2.31) or an estimate of the transmissivity

$$T_s = K_s \times \Delta z = C_p \times H \times T_1^2 \tag{7.1}$$

where $K_s$ is the saturated hydraulic conductivity, $\Delta z$ is the thickness of the aquifer formation and $H$ is the equivalent water column.

The water column estimated using the MRS inverse model is a more stable parameter than the water content or the thickness of a water-saturated layer. It makes the transmissivity a more reliable estimator than the hydraulic conductivity. The transmissivity estimator (equation (7.1)) was tested using MRS measurements in France. The test area between Chartres and Orleans (figure 7.2) makes part of the Loire basin belonging to the Paris Basin. The subsurface is composed of intercalations of clay, marl, sand and limestone or chalk formations. Limestone and chalk may be partly fractured with a local karst development. For computing $T_s$, the empirical constant $C_p$ is selected so that the MRS estimated transmissivities match the best transmissivities measured with pumping tests. The test sites in the same geological environment require only one value of $C_p$.

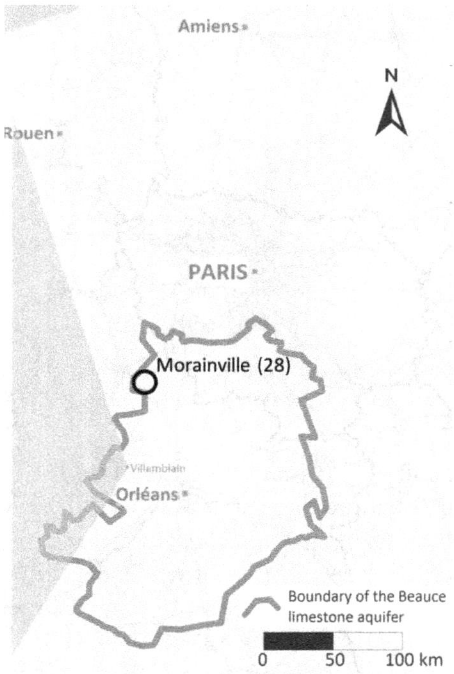


**Figure 7.2.** Location map of the investigated with MRS area in France.

Figure 7.3 shows an example of the comparison between the transmissivity estimated with MRS and that given by borehole pumping tests.

Equation (7.1) shows that the transmissivity estimate is sensitive to the accuracy of the $T_1$ measurements. For measuring $T_1$ corresponding to one depth interval, a number ($I_d$) of two-pulse measurements with equal pulse moments but different delays between the pulses is carried out. The measurement of $T_1$ versus depth needs to perform a sounding by varying the pulse moment. An MRS sounding comprises measurements with $I$ different values of the pulse moment and thus, an accurate $T_1$ sounding needs $I \times I_d$ measurements. One typical MRS sounding with seize pulse moments takes between one and ten hours and thus, an accurate $T_1$ sounding being $I_d$ times longer is a time consuming procedure. A practically acceptable procedure for measuring $T_1$ comprises using only one-delay value per sounding. The second pulse is generated immediately after the FID1 signal and the FID2 signal is measured after the second pulse. This one-delay procedure is a compromise between the accuracy of measuring $T_1$ and the measuring time. An MRS inversion for $T_1$ is done using three values: the FID1 amplitude measured after the first pulse, the FID2 amplitude measured after the second pulse and the assumption that $v_{02}(\tau_d = 0) = 0$. However, this assumption is not always respected when investigating real geological formations and because of technical difficulties of accurately generating two equal pulses.

Figure 7.4 shows the FID2 amplitudes computed considering the zero-shift (black crosses) and a non-zero-shift (blue circles). Both signals have equal FID1 amplitudes and equal $T_1$ values. The vertical dashed line corresponds to the delay value used for measuring. These graphs show a higher measured amplitude caused by the non-zero-shift. When fitted with the zero-shift assumption, which is true only for one of them (black crosses), the non-zero-shift causes an underestimate of $T_1$. This error can be damped at a reasonable price by performing additional $T_1$ measurements by varying the delay but using only one pulse moment value corresponding to the maximum FID1 amplitude.

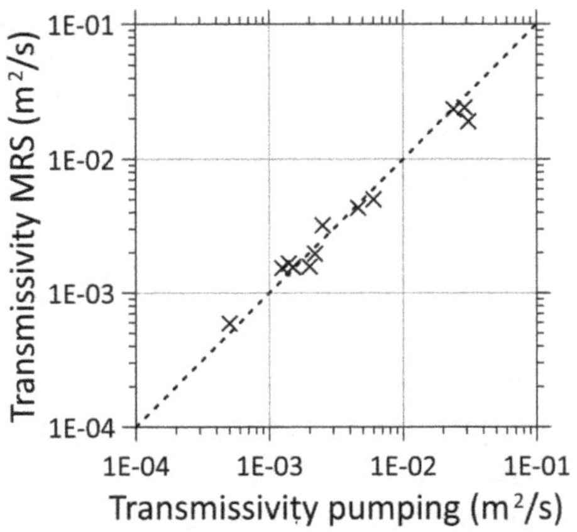


**Figure 7.3.** Example of the transmissivity estimated with MRS versus that given by pumping tests. The dashed line shows the perfect correlation.

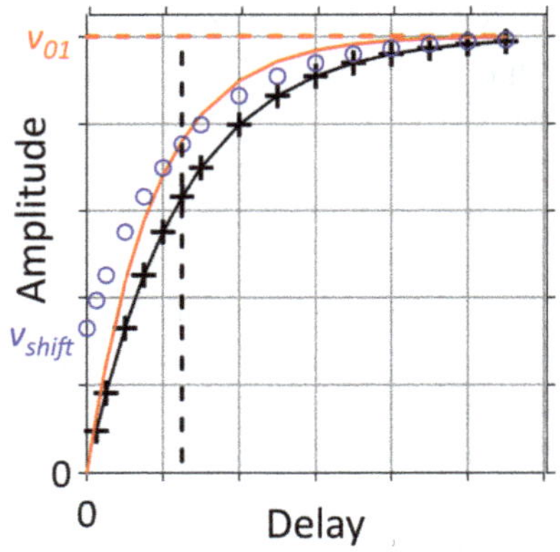


**Figure 7.4.** Amplitude of the FID2 signal versus delay between the pulses. Black crosses show the amplitude computed with $v_{\text{shift}} = 0$ and circles that computed considering $v_{\text{shift}}>0$. The black dashed line shows the delay value for measuring, the red dashed line shows the FID1 amplitude and the solid lines show exponential signals computed without $v_{\text{shift}}$ (black) and with $v_{\text{shift}}$ (red).

The one-delay sounding procedure is sensitive to the accuracy of measurements and hence experimental errors may degrade the accuracy of the transmissivity estimate. For this reason, the one-delay $T_1$ inversion is strongly regularized and provides only one reliable $T_1$ value corresponding to the maximum of the MRS amplitude. These results can be improved only by additional measurements.

## 7.2 Field example: processing one MRS sounding

Now, we demonstrate a step-by-step interpretation of one MRS sounding using field data measured in the Morainville test site (figure 7.2). In the shallow part, the subsurface is composed of intercalated sand, clay and limestone layers (figure 7.5). The sand aquifer between 38 and 68 m is confined.

Under Morainville measuring conditions, the maximum depth of water detection is estimated at approximately 100 m. Figure 7.5 shows results of the regularized MRS inversion carried out using a 12-layer model (red line), which corresponds to the maximum theoretical resolution ($\mathbf{R_m} = \mathbf{I}$). Figure 7.6 shows the 95% confidence interval for this inverse model estimated with the SVD method. Results show that shallow and deep layers can be resolved with a smaller accuracy than layers between 10 and 50 m. The SVD estimate allows selecting the mean value for the Monte Carlo simulation $\Delta w_{n,j} = 0.06$.

The inversion for water content and the one-delay inversion for $T_1$ fit well the experimental data (figure 7.7). The equivalent inverse models ($H_{\min}$ and $H_{\max}$) fit the measured signals equally well and are not shown in figure 7.7.

Results of the multi-layer inversion show two water-bearing formations and suggest testing a 2-layer inversion. We compute three 2-layer inverse models with different water contents. The theoretical signals corresponding to these models fit measured amplitudes with about equal accuracy. Figure 7.8 shows the FID1

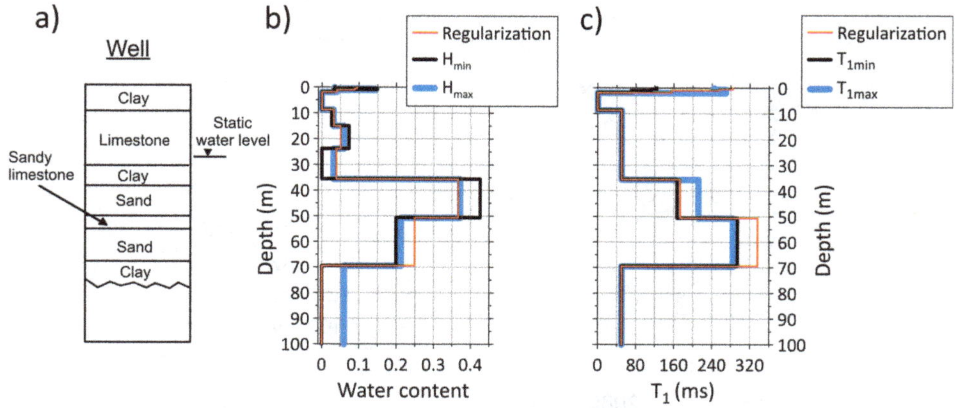


**Figure 7.5.** MRS study in the Morainville site: (a) the lithological log of borehole F1; (b) the MRS water content versus depth given by three equivalent inverse models; (c) the relaxation time $T_1$ versus depth given by the regularized inverse model (red line) and its min–max values.

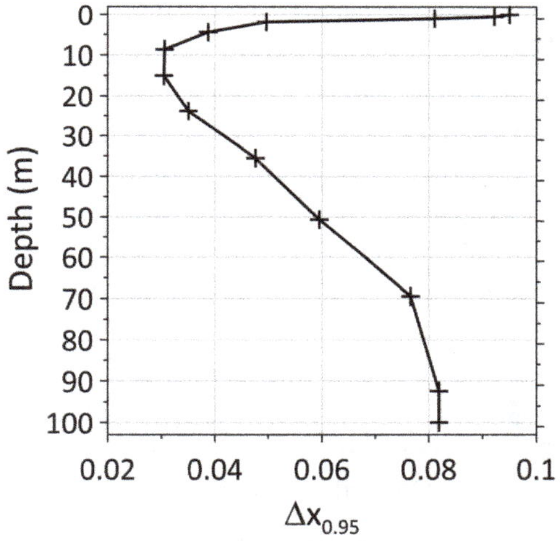


**Figure 7.6.** MRS study in the Morainville site: the 95% confidence interval for the discretization with 12 layers.

amplitude (crosses) and the inversion fits. FID2 amplitudes are fitted with the accuracy equal to that for the regularized solution. A 1-layer model does not allow fitting measured amplitudes with the acceptable accuracy.

Figure 7.9 shows the borehole and three 2-layer equivalent models.

In total, we got six equivalent inverse models and the max–min estimates of the relaxation time $T_1$ for each model (table 7.2).

The MRS inverse models show the aquifer between 30 and 70 m. In this aquifer, the water content ranges between the saturated soil water content $\theta_s$ (typical values for medium sand 0.4–0.43) and the specific yield (typical values 0.15–0.32). Between

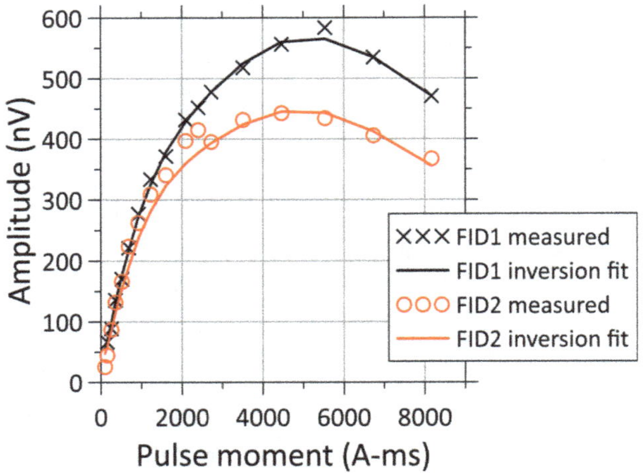


**Figure 7.7.** MRS study in the Morainville site: measured FID1 and FID2 signals versus pulse moment (symbols) and the theoretical signals computed after inversion with 12 model layers (solid lines).

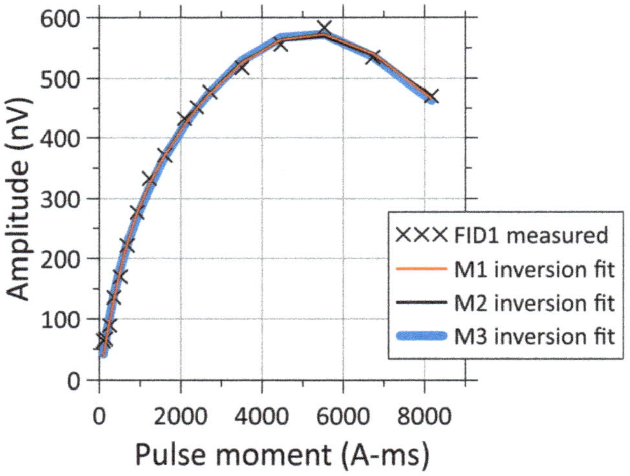


**Figure 7.8.** MRS study in the Morainville site: measured FID1 amplitude versus pulse moment (crosses) and the theoretical signals computed after inversion with three 2-layer inverse models (solid lines).

10 and 30 m, MRS shows water in the unsaturated limestone ($w \approx 0.05$). Only one model ($H_{\text{max}}$) suggests water below 70 m ($w \approx 0.06$). Thus, MRS does not exclude some little water below 70 m, but the MRS resolution does not allow for resolving this formation. The borehole shows clay below 70 m.

The transmissivity estimate provided by the inverse models (table 7.3) ranges between 2.5E-3 and 4.9E-3 $\text{m}^2\ \text{s}^{-1}$ with the mean value of 3.79E-3 $\text{m}^2\ \text{s}^{-1}$. Different interpretations of the pumping test report the transmissivity between 2.2E-3 and 6.0E-3 $\text{m}^2\ \text{s}^{-1}$ with the mean value of 4.1E-3 $\text{m}^2\ \text{s}^{-1}$.

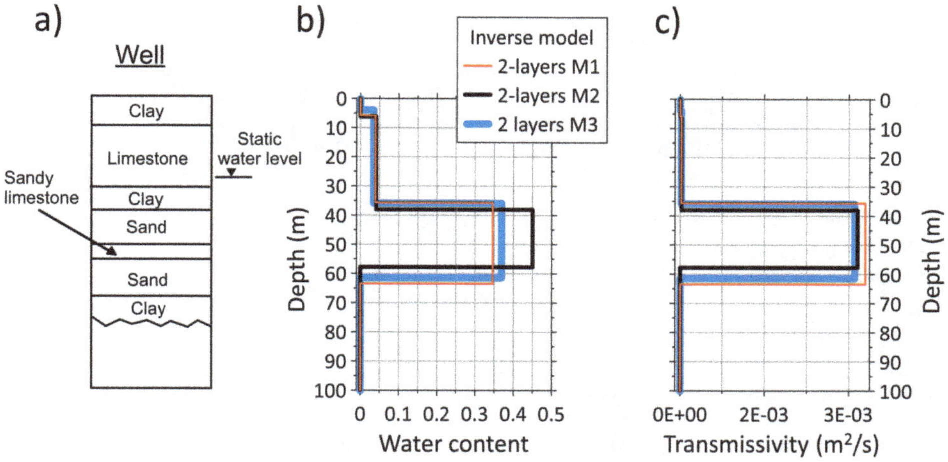


**Figure 7.9.** MRS study in the Morainville site: (a) the lithological log of borehole *F*1; (b) the MRS water content versus depth given by three equivalent 2-layer inverse models; (c) the relaxation time $T_1$ versus depth given by these inverse models.

**Table 7.2.** MRS study in the Morainville site: the inverse models provided by the MRS inversion.

| Number | Inverse model | |
|---|---|---|
| 1 | 12-layer regularization | Relax. time $T1_{min}$ |
| 2 | 12-layer $H_{min}$ | Relax. time $T1_{min}$ |
| 3 | 12-layer $H_{max}$ | Relax. time $T1_{min}$ |
| 4 | 2-layer *M*1 | Relax. time $T1_{min}$ |
| 5 | 2-layer *M*2 | Relax. time $T1_{min}$ |
| 6 | 2-layer *M*3 | Relax. time $T1_{min}$ |
| 7 | 12-layer regularization | Relax. time *T*1 |
| 8 | 12-layer $H_{min}$ | Relax. time *T*1 |
| 9 | 12-layer $H_{max}$ | Relax. time *T*1 |
| 10 | 2-layer *M*1 | Relax. time *T*1 |
| 11 | 2-layer *M*2 | Relax. time *T*1 |
| 12 | 2-layer *M*3 | Relax. time *T*1 |
| 13 | 12-layer regularization | Relax. time $T1_{max}$ |
| 14 | 12-layer $H_{min}$ | Relax. time $T1_{max}$ |
| 15 | 12-layer $H_{max}$ | Relax. time $T1_{max}$ |
| 16 | 2-layer *M*1 | Relax. time $T1_{max}$ |
| 17 | 2-layer *M*2 | Relax. time $T1_{max}$ |
| 18 | 2-layer *M*3 | Relax. time $T1_{max}$ |

Table 7.4 summarizes the equivalent water column computed considering the depth interval between 8 and 100 m and the fitting error for each inverse model. The fitting error varies between 5.99 and 10.35 nV, which is less than the noise measured at 12.69 nV.

Figure 7.10 visualizes results presented in tables 7.2–7.4.

**Table 7.3.** MRS study in the Morainville site: the transmissivity estimated with different inverse models and that with pumping tests in the borehole.

| | Transmissivity ($m^2\ s^{-1}$) | | | | |
|---|---|---|---|---|---|
| Inverse model | Relax. time $T1_{min}$ | Relax. time $T1$ | Relax. time $T1_{max}$ | Pumping min | Pumping max |
| 12-layer regularization | 3.94E-03 | 4.87E-03 | 4.40E-03 | | |
| 12-layer $H_{min}$ | 3.58E-03 | 4.65E-03 | 4.15E-03 | | |
| 12-layer $H_{max}$ | 3.55E-03 | 4.68E-03 | 4.03E-03 | | |
| 2-layer $M1$ | 2.72E-03 | 3.54E-03 | 4.23E-03 | | |
| 2-layer $M2$ | 2.51E-03 | 3.39E-03 | 3.91E-03 | | |
| 2-layer $M3$ | 2.66E-03 | 3.33E-03 | 4.15E-03 | | |
| Borehole | | | | 2.20E-03 | 6.00E-03 |

**Table 7.4.** MRS study in the Morainville site: the water column estimated with different inverse models and the fitting error corresponding to each inverse model.

| Number | Inverse model | $H$ (m) | RMSE (nV) |
|---|---|---|---|
| 1 | 12-layer regularization | 11.30 | 6.75 |
| 2 | 12-layer $H_{min}$ | 10.98 | 5.99 |
| 3 | 12-layer $H_{max}$ | 13.71 | 6.9 |
| 4 | 2-layer $M1$ | 10.79 | 9.1 |
| 5 | 2-layer $M2$ | 10.20 | 8.93 |
| 6 | 2-layer $M3$ | 10.54 | 10.35 |
| Mean $H$ | | 11.25 | — |
| Noise (nV) | | | 12.69 |

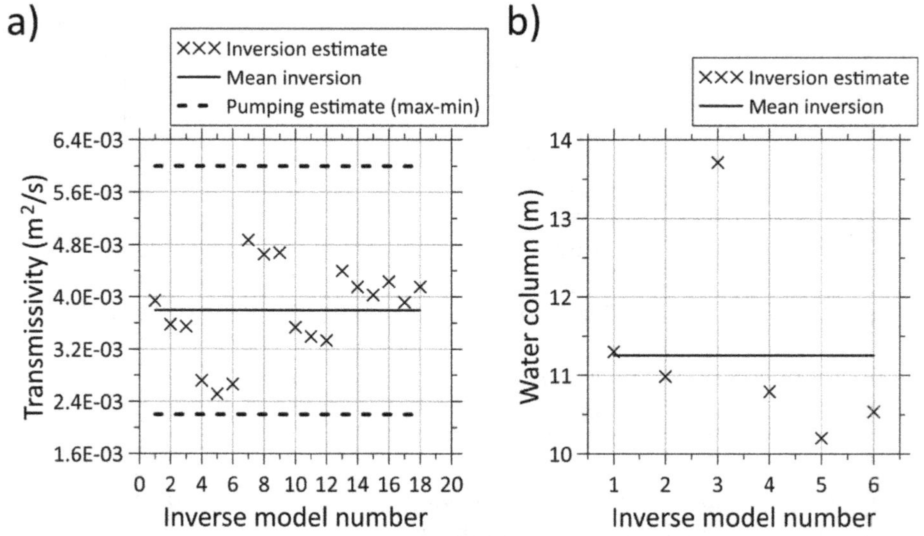


**Figure 7.10.** MRS study in the Morainville site. (a) The transmissivity estimated with MRS (crosses) versus inverse model number with the mean shown by the solid line (the mean = 3.79E-3 $m^2\ s^{-1}$). The dashed lines show the maximum and minimum values of the transmissivity provided by interpretation of the pumping tests (the mean = 4.1E-3 $m^2\ s^{-1}$). (b) The water column estimated with MRS (crosses) versus inverse model number with the mean shown by the solid line.

Note that an analysis of the MRS results allows for estimating the uncertainty in the MRS inverse model. The correspondence between the MRS results and the investigated subsurface assumes using the ground truth. For that, pumping test and logging in a borehole can be considered.

## 7.3 Field example: mapping a catchment

Mapping at the scale of a catchment is the typical application for geophysical methods, including MRS. An area investigated with one MRS station is approximately $100 \times 100\ m^2$ and a catchment with a large surface area can be investigated by a set of 1D soundings. Mapping with a 2D or a 3D MRS setup is used in a karst environment or investigating a fractured zone in hard rocks. The presented field study of the Kerbernez catchment shows an example of MRS application to mapping aquifers at a catchment scale. The data processing and inversion are not different from that discussed in detail in the previous chapters and the goal is to show what results MRS can provide for hydrogeology. Thus, this study can be considered as an example of a fieldwork report with MRS.

The Kerbernez catchment ($0.1\ km^2$) is located about 10 km south of Quimper, France (figure 7.11).

The aquifer is unconfined and composed of fractured granite covered by a few meters of sandy–clayey weathered material. Around the area, the thickness and hydraulic conductivity of the weathered part of the aquifer both vary significantly.

**Figure 7.11.** Kerbernez test site location map.

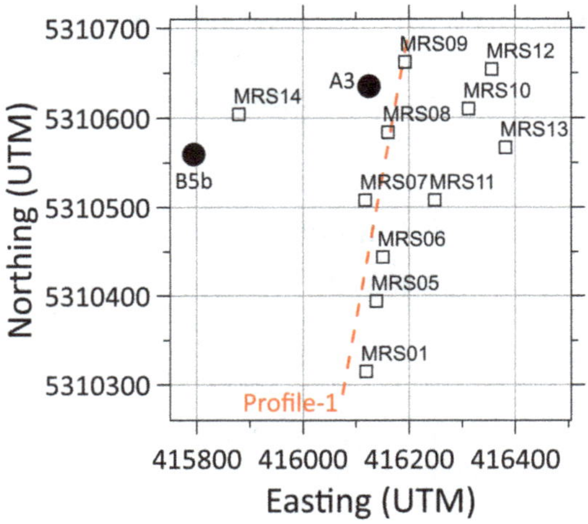


**Figure 7.12.** Kerbernez test site: position of MRS stations, MRS Profile-1 and two boreholes.

Eleven MRSs were carried out using the "NUMIS$^{PLUS}$" system manufactured by IRIS Instruments, and boreholes were available to validate the MRS results (figure 7.12).

For measuring, a noise-reducing figure-eight 37.5-m-side loop was used for compensating a strong electromagnetic noise observed during fieldwork. With this loop, the depth of investigation is estimated at approximately 40 m, and the investigated subsurface can be approximated by a box of $\approx 60 \times 120 \times 40$ m$^3$. We estimate the hydraulic conductivity of weathered granite aquifers using the empirical constant developed for sand aquifers ($C_p = 7.0 \times 10^{-9}$). However, the MRS water content could not be calibrated because no reliable measurements of the water content are available.

We observe strong variations of the MRS signal amplitude pointing out that the subsurface is heterogeneous. Figure 7.13 shows amplitudes versus pulse moment measured at two MRS stations with different MRS signals. The electromagnetic noise (black squares) allows for estimating the signal to noise ratio.

At all MRS stations, the amplitude measured versus pulse moment has a pronounced maximum and a map of the MRS signal amplitude summarizes observed subsurface variations before inversion (figure 7.14). Such a map is always recommended to draw because it provides an overview not linked with the forward modeling and inversion. Figure 7.14 shows higher amplitudes in the south–eastern part of the catchment, thus pointing to the aquifer location.

In figures 7.15 and 7.16, one can see the inverse models corresponding to two MRS stations.

The MRS log near borehole A3 (figure 7.15) shows that both the water content and the relaxation time increase from approximately 3 to 15 m. This depth interval is in good agreement with the aquifer, shown by the borehole. The MRS water content suggests that the depth to the top of the aquifer is between 4 and 7 m, which is consistent with the static water level measured in the borehole at 6.04 m. This large uncertainty in MRS results can be explained by different factors: the equivalence

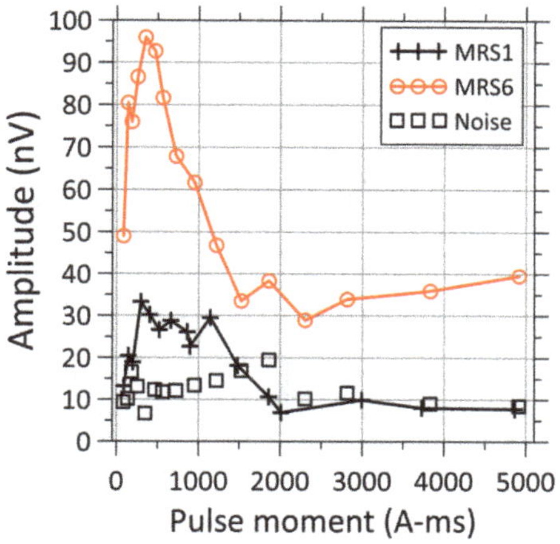


**Figure 7.13.** Kerbernez test site: example of the amplitude versus pulse moment measured at two MRS stations. The noise amplitude shows the minimum detectable MRS signal.

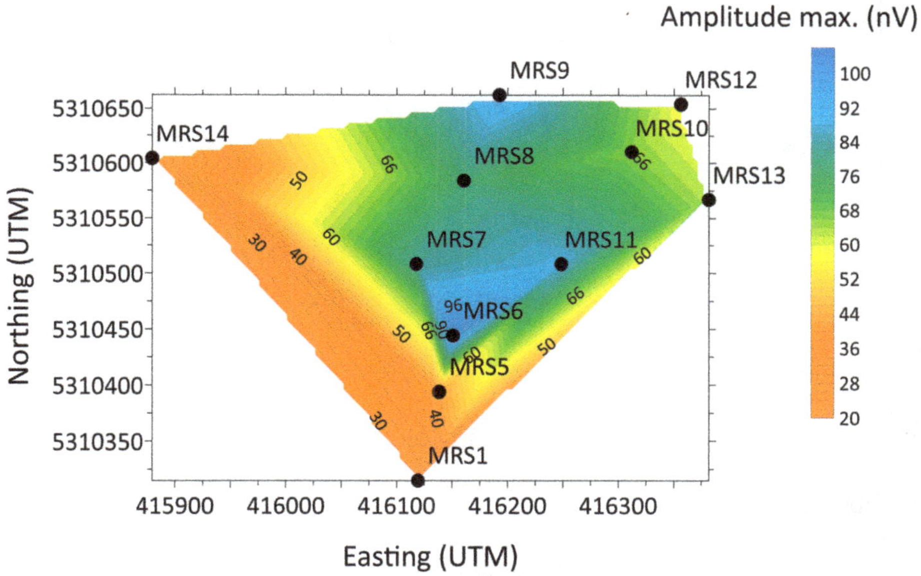


**Figure 7.14.** Kerbernez test site: a map of the maximum amplitude of the MRS signal.

problem; 1D measurements over the heterogeneous subsurface; capillary water above the saturated zone, where the water content may change smoothly from the static water level towards the surface. Pumping tests carried out in borehole A3 reveal a transmissivity $T_{bh} = 4.4 \times 10^{-4}$ m$^2$ s$^{-1}$, and the MRS estimated transmissivity is $T_{MRS} = (3.7 \pm 1.8) \times 10^{-4}$ m$^2$ s$^{-1}$.

Borehole B5b (figure 7.16) reveals a low-permeable clay material and, for this reason, no pumping tests were carried out. MRS results derived from the MRS14 sounding show a three times smaller water content and four times shorter relaxation

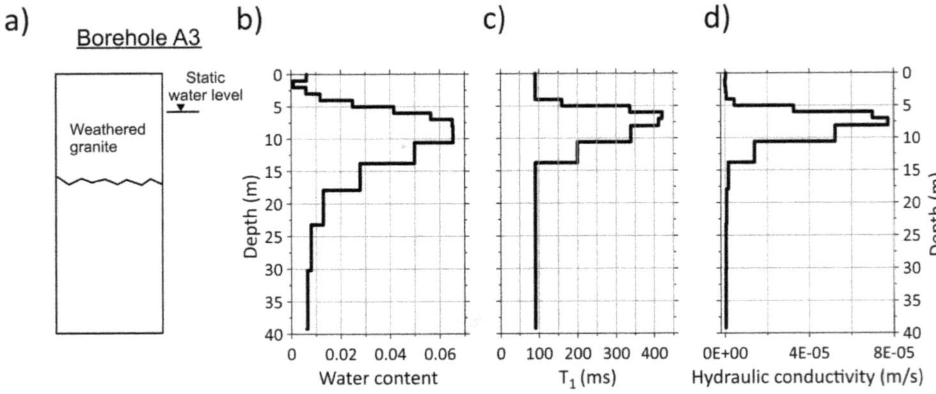


**Figure 7.15.** Kerbernez test site: example of MRS results (MRS9 station) near borehole A3.

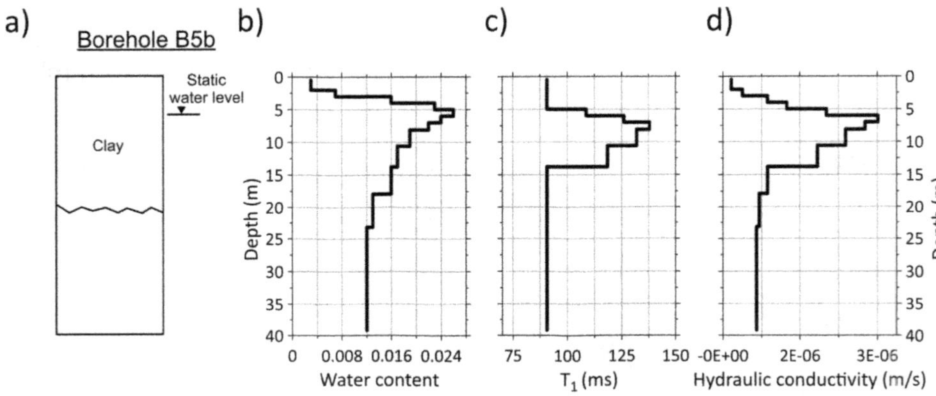


**Figure 7.16.** Kerbernez test site: example of MRS results (MRS14 station) near borehole B5b.

time than that observed at the MRS9 station (weathered granite). Thus, MRS14 sounding points to a low-permeable subsurface. The static water level measured in the borehole (5.48 m) corresponds to the maximum MRS water content. Thus, a good qualitative agreement between the MRS and borehole results is observed. Note that the one-delay sounding was not configured for measuring short relaxation times and the relaxation time $T_1$ below 15 m shown at 90 ms is given by the one-delay inversion with a low accuracy. Most probably, $T_1$ is overestimated and the true value is smaller than that shown by the MRS inverse model. MRS shows the water content value of 0.01 between 15 and 40 m. However, this value may be overestimated because of the noise voltage that imposes a threshold for detection of small signals: the MRS signal smaller than the noise cannot be identified. Note that when the MRS signal is smaller than noise, it would be erroneous to say that MRS does not work. It does and shows that the water amount is small, which may be valuable information for a hydrogeological study.

Figure 7.17 summarizes the MRS results along Profile-1 (figure 7.12).

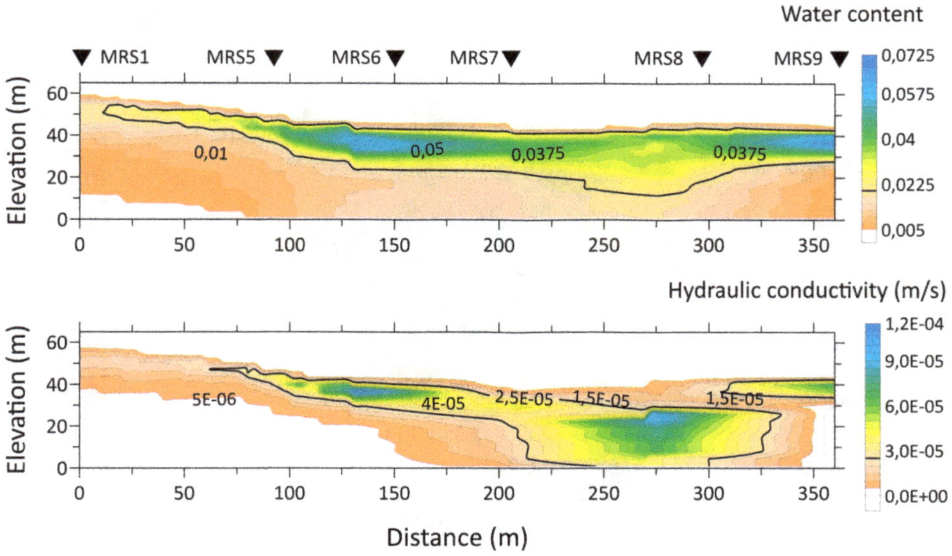


**Figure 7.17.** Kerbernez test site: MRS results along Profile-1.

The MRS data delineate well the aquifer and show strong variations in both the water content and the hydraulic conductivity. In the northern part of Profile-1, between stations MRS1 and MRS5, the subsurface is characterized by a water content of about 0.01 and a hydraulic conductivity of less than $10^{-5}$ m $s^{-1}$. In geological terms, these MRS results may be interpreted as either a fissured or solid granite or a clay formation. In the southern part of the profile, rocks correspond to water-saturated weathered granite or sand. Thus, the main aquifer is located downhill in a weathered zone corresponding to a deep graben structure that acts as a water storage for this catchment.

Figures 7.18 and 7.19 finalize the MRS results by showing the maps of the equivalent water column and of the transmissivity corresponding to the water storage and the water flow parameters of the aquifer, respectively.

## 7.4 Field example: MRS contribution to an unsaturated water flow modeling

The selective sensitivity of the MRS method to groundwater renders it attractive for using in hydrogeological modeling. The very first idea was to apply the method to noninvasive measurements of the groundwater water level (GWL). However, measuring the GWL is possible only for unconfined aquifers and when the aquifer is confined, the MRS shows the top of the aquifer, which may not coincide with the GWL. The accuracy of the MRS inversion allows measuring the aquifers geometry and hence the GWL with the uncertainty of more than a few meters. This poor accuracy in not acceptable for hydrogeological modeling. The principal result of the MRS measurement is an estimate of the aquifer effective porosity and the saturated hydraulic conductivity. These parameters can be successfully used for constraining hydrogeological models and render

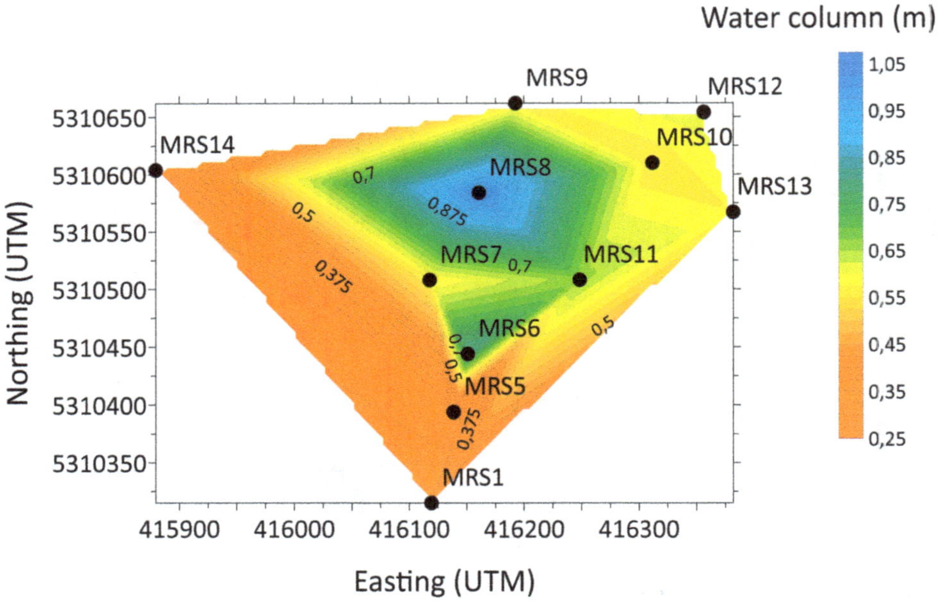


**Figure 7.18.** Kerbernez test site: a map of the water column estimate provided by the MRS inversion for water content.

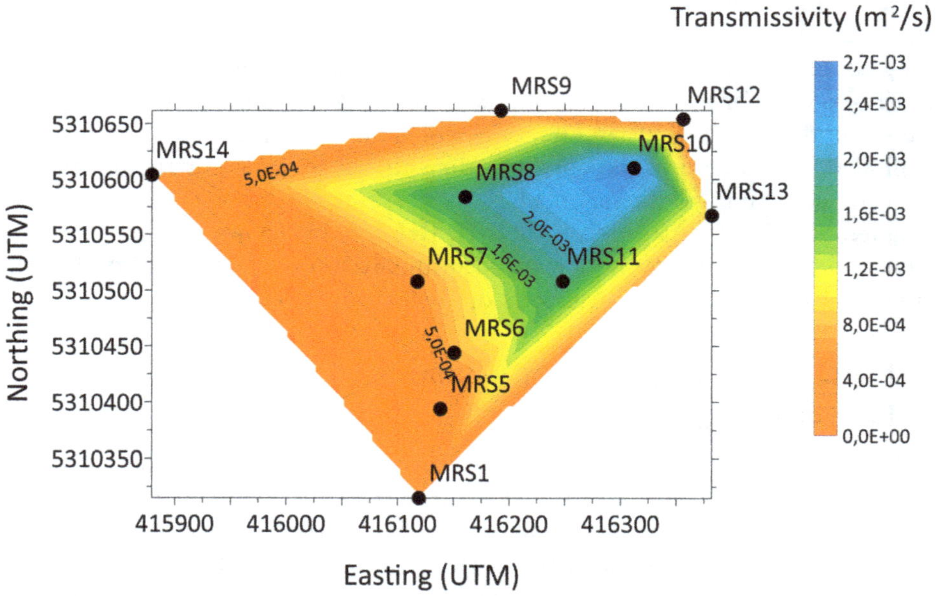


**Figure 7.19.** Kerbernez test site: a map of the transmissivity estimate provided by the MRS inversion.

the model closer to a real hydrological system. In the water-saturated media, the aquifer effective porosity and the saturated hydraulic conductivity are invariable in time. Variations of the GWL only affect the top of the first formation and except in some particular cases, time-lapse monitoring is unnecessary.

In the unsaturated porous media, the hydraulic conductivity is a non-linear function of the water content and the estimate of the unsaturated hydraulic conductivity using the procedure developed for water-saturated porous media provides acceptable results only when the unsaturated media is close to saturation. This reduces the importance of the relaxation times for characterizing water in the unsaturated zone. In a nonmagnetic geological material like chalk or limestone, the water content can be measured with the MRS in both the saturated and unsaturated zones. The water content may vary in time and time-lapse MRS measurements can be a power tool for monitoring water content variations in the unsaturated zone.

We investigate the capacity of the MRS 1D inversion to delineate water in the saturated and unsaturated zones. For that, we perform a numerical modeling of the MRS response using six models presented in table 7.5. Each model comprises a layer between 20 and 50 m simulating the saturated zone and three 5-m-thick layers between 5 and 20 m simulating the unsaturated zone. We assume the water content in the saturated zone large ($w = 0.25$) or small ($w = 0.05$). The water content in the unsaturated zone is represented by three layers with $w_1 = 0.05$, $w_2 = 0.15$ and $w_3 = 0.2$ for the models with a large water column $H_{\text{unsat}}$ and with $w_1 = 0.02$, $w_2 = 0.08$ and $w_3 = 0.1$ for the models with a medium water column. Two models assume no water in the unsaturated zone. These models allow investigating different scenarios that comprise a variable ratio between the water quantity in the saturated and unsaturated zones. Inversion is carried out using the Tikhonov regularization with a 15-layer inverse model selected considering the ideal resolution condition $\mathbf{R}_m = \mathbf{I}$. The water column is calculated in the unsaturated ($H_{\text{unsat}}$) and saturated ($H_{\text{sat}}$) zones. The relative error of the water column estimate is computed relative to the true value given by the model.

Figure 7.20 shows the amplitude of the MRS signal versus pulse moment computed considering the models with a lot of water between 20 and 50 m. For rendering our modeling more realistic, we add a random noise of approximately 10 nV which is often observed in the field. One can see that water in the unsaturated zone causes a small increase of the amplitude. Processing data with such a small difference requires MRS measurement of a good accuracy.

Inversion of these data sets (figure 7.21) shows that all three models are well resolved by inversion. The inverse models show well variations of the water content corresponding to the unsaturated zone above 20 m.

**Table 7.5.** Modeling MRS estimate of the water column in the unsaturated zone varying the ratio of water amount in the unsaturated and saturated zones.

| Model | $H_{\text{mod}}$ (m) | $H_{\text{inv}}$ (m) | Error (%) | Model description |
|---|---|---|---|---|
| Model-1 | 2.00 | 1.75 | −12.36 | $H_{\text{unsat}}$ large, $H_{\text{sat}}$ large |
| Model-2 | 1.00 | 0.87 | −12.96 | $H_{\text{unsat}}$ medium, $H_{\text{sat}}$ large |
| Model-3 | 1.00 | 0.85 | −14.56 | $H_{\text{unsat}}$ medium, $H_{\text{sat}}$ small |
| Model-4 | 2.00 | 1.55 | −22.68 | $H_{\text{unsat}}$ large, $H_{\text{sat}}$ small |
| Model-5 | 0.00 | 0.18 | | $H_{\text{unsat}}$ =0, $H_{\text{sat}}$ large |
| Model-6 | 0.00 | 0.05 | | $H_{\text{unsat}}$ =0, $H_{\text{sat}}$ small |

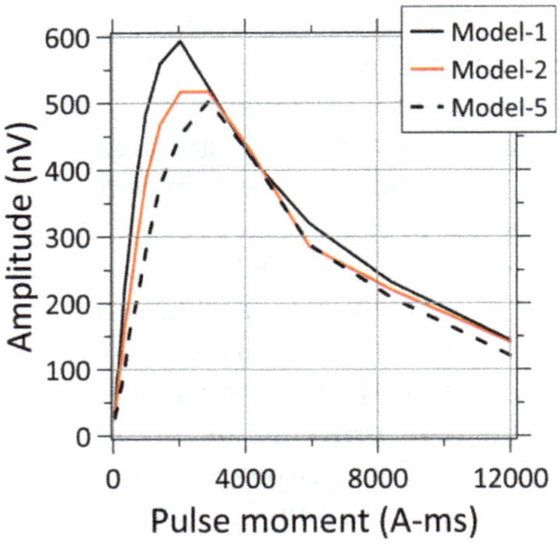


**Figure 7.20.** MRS amplitude versus pulse moment computed considering the models that comprise a lot of water in the saturated zone.

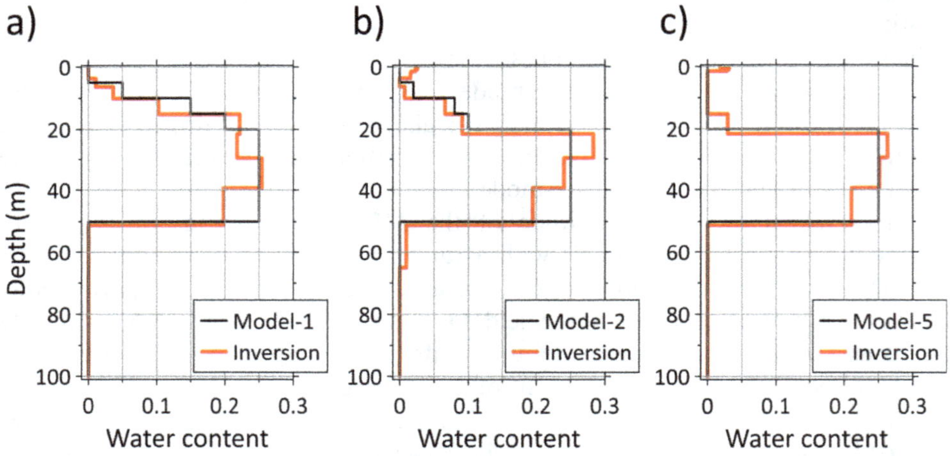


**Figure 7.21.** MRS inversion for water content (red lines) considering different models containing a lot of water in the saturated zone (black lines): (a) a lot of water in the unsaturated zones; (b) less water in the unsaturated zone; (c) no water in the unsaturated zone.

Synthetic signals (figure 7.22) and the inversion results (figure 7.23) corresponding to the models with a small amount of water below 20 m show that contribution of water in the unsaturated zone is significant and can be easily detected in raw data. Inversion shows the corresponding difference in the water content above 20 m.

The water column estimate with the MRS inversion plotted versus the true value given by the initial model (figure 7.24) corresponds to the accuracy of the water column estimate presented in table 7.5.

We present an example of the MRS application to investigation of water content variations in the unsaturated zone observed in the Villamblain test site in France

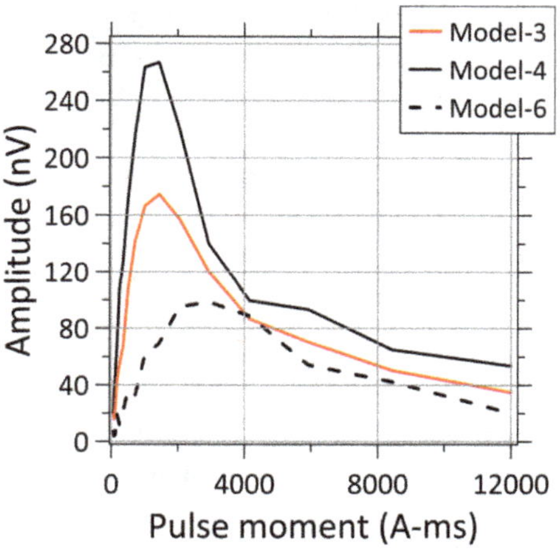


**Figure 7.22.** MRS amplitude versus pulse moment computed considering the models that comprise a small amount of water in the saturated zone.

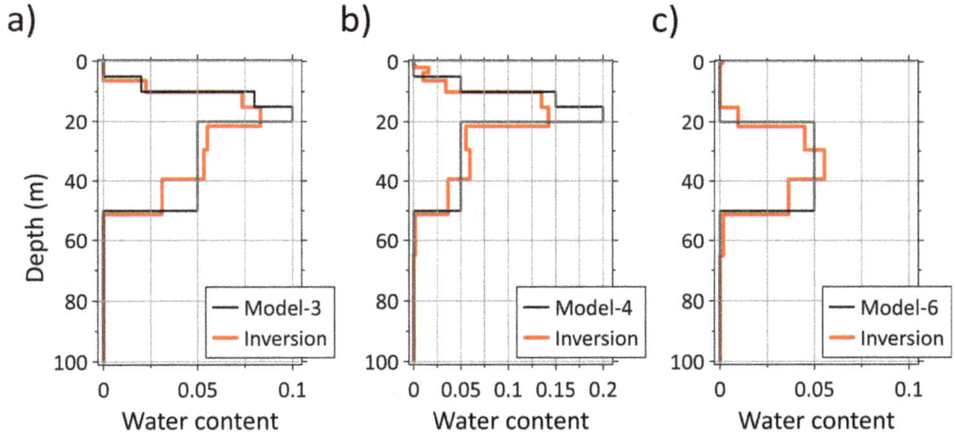


**Figure 7.23.** MRS inversion for water content (red lines) considering different models containing a small amount of water in the saturated zone (black lines): (a) a lot of water in the unsaturated zones; (b) less water in the unsaturated zone; (c) no water in the unsaturated zone.

(figure 7.25). The Villamblain test site covers about 7 km$^2$. This area is a part of the Beauce aquifer (9000 km$^2$) located south-west of Paris. The mean annual recharge varies around 110 mm. In Villamblain, soil thickness ranges from 0.3 to 1.5 m. The water column in these soils varies between 50 and 180 mm. The limestone underlying the soil layer is heterogeneous with the variable fracturing and irregular karst development. Estimations of the specific yield of the limestone show 3%–13%. All the studies show no-run-off in Villamblain.

During our study, we measured the GWL in the borehole 03622X0119/FP3-25. The Villampuy weather station at 4 km from Villamblain (Lat.-48°02'18"N; Long.-1°29'30"E) provides the daily rainfall data. We calculate the effective rainfall (ERF)

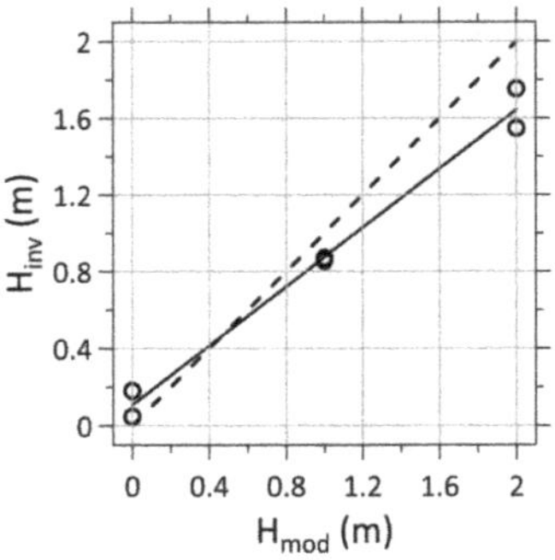


**Figure 7.24.** The water column in the unsaturated zone estimated by MRS inversion versus the true value given by the model (circles). The solid line shows a linear fit and the dashed line shows the perfect correspondence.

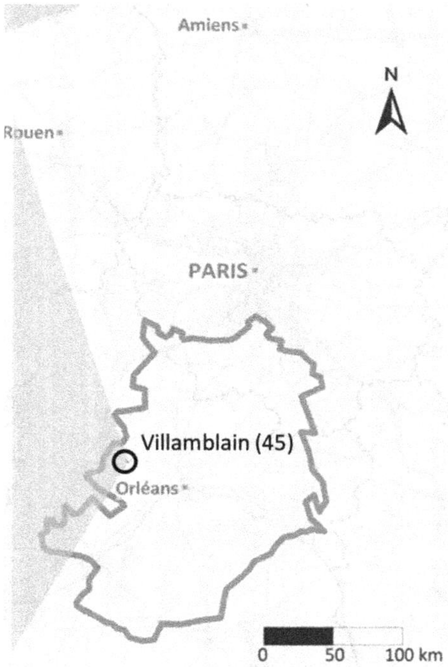


**Figure 7.25.** Villamblain test site: location map of the monitoring site.

as a difference between the rainfall (RF) and the actual evapotranspiration (AET). We have no data for calculating AET, and we simplify the estimate using the Penman potential evapotranspiration (PET)

$$\begin{cases} \text{ERF} = \text{RF} - \text{PET}, & \text{if} \quad \text{RF} > \text{PET} \\ \text{ERF} = 0, & \text{if} \quad \text{RF} \leqslant \text{PET} \end{cases}. \tag{7.2}$$

A one-year monitoring was performed between 26 April 1999 and 15 March 2000. The data set comprises 34 MRS soundings at 18 different dates. We used the

NUMIS instrument with a 75 × 75 $m^2$ square loop and the maximum pulse moment of 12 500 A-ms. The Larmor frequency was 2011.3 Hz. Figure 7.26 shows the amplitude of the MRS signal corresponding to two soundings with the maximum and the minimum amplitudes. Squares show the typical electromagnetic noise after stacking.

The MRS signal has a pronounced maximum corresponding to the pulse moment of about 1700 A-ms. The maximum of the MRS signal corresponding to each sounding is plotted versus the day of monitoring (figure 7.27). This plot shows increasing amplitudes after 175 days. This increase in fall 1999 approximately corresponds to the beginning of the winter recharge of the aquifer system.

Comparison of the MRS water content log (figure 7.28(b)) with the lithological log of the nearby borehole (figure 7.28(a)) suggests that in the limestone geological environment, MRS detects water in the unsaturated zone. The MRS log shows a smooth increase of the water content from the surface to the GWL. Between 4 and 8 m, the MRS reveals a low water content, suggesting a clay layer observed at this depth in the borehole. The observed low water content may also correspond to an unsaturated fractured limestone that does not contain much water. Because of the limited resolution, the MRS inversion did not identify the GWL at 17 m. Below 25 m, decreasing water content corresponds to a lowering specific yield of the saturated limestone.

For this sounding, the theoretical signal fits the measured one well (figure 7.29).

Figure 7.30 shows results of the time-lapse MRS inversion. The water content (color scale) corresponding to each sounding (monitoring day) is plotted versus depth. The MRS inverse model identifies two principal water storage zones between 0 and 2 m and below 10 m. The water content in limestone ranges between 0.04 and 0.08. These values are in good agreement with the specific yield in limestone reported

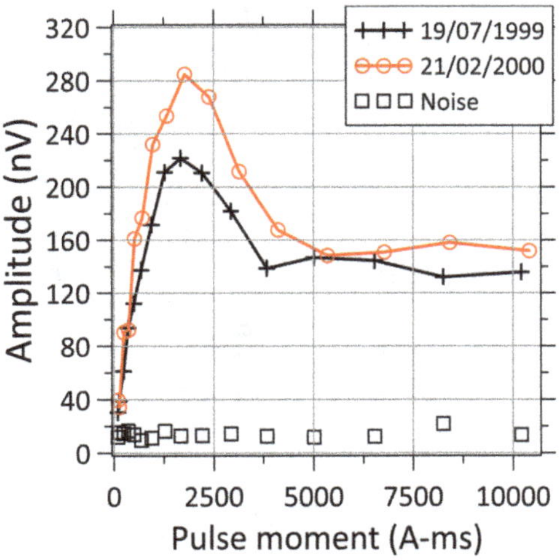


**Figure 7.26.** Villamblain test site: example of the MRS amplitude versus pulse moment corresponding to the maximum and the minimum MRS signals. Black squares show the typical electromagnetic noise measured during this study.

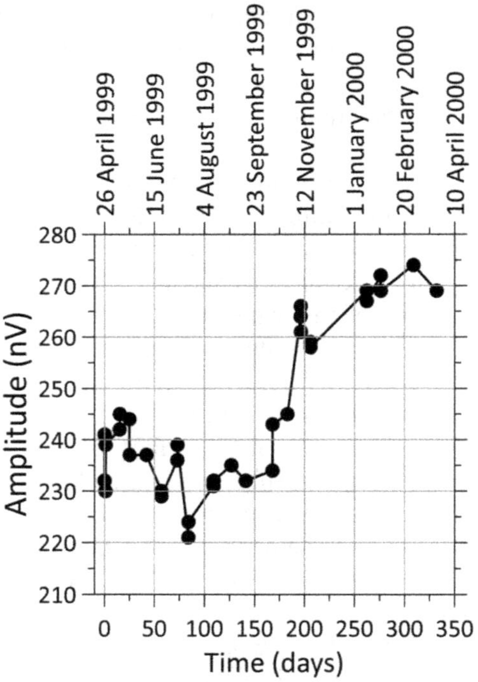


**Figure 7.27.** Villamblain test site: the maximum amplitude of the MRS signal corresponding to each MRS sounding performed during the study.

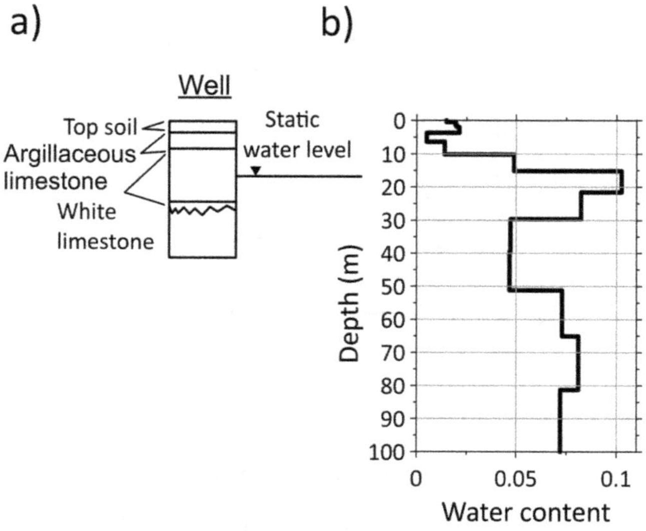


**Figure 7.28.** Villamblain test site: (a) the lithological log of the borehole 03622X0119/FP3-25 located near the MRS station; (b) the water content provided by the MRS inverse model.

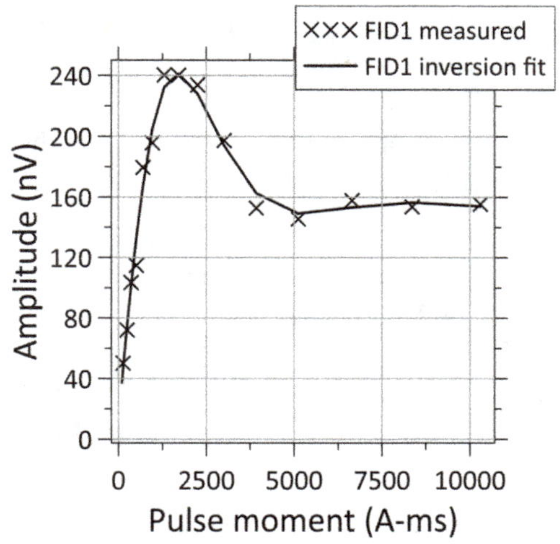


**Figure 7.29.** Villamblain test site: the amplitude of the MRS signal versus pulse moment (crosses) and the theoretical amplitude computed after the MRS inverse model.

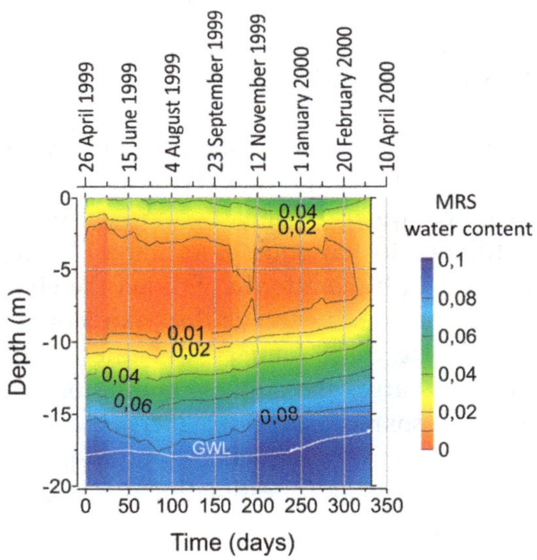


**Figure 7.30.** Villamblain test site: the water content provided by the time-lapse MRS inversion.

between 0.03 and 0.13. The MRS reveals less water between 2 and 10 m. This observation may have two plausible explanations. The first one suggests a low-permeable clay formation. The second explanation puts forward a permeable formation that does not store much water when unsaturated (gravel or karst). MRS alone cannot distinguish between these two possibilities.

Figure 7.31(a) shows the water column measured with MRS in the depth interval between 0 and 18 m (black circles) and the red line shows the cumulative effective

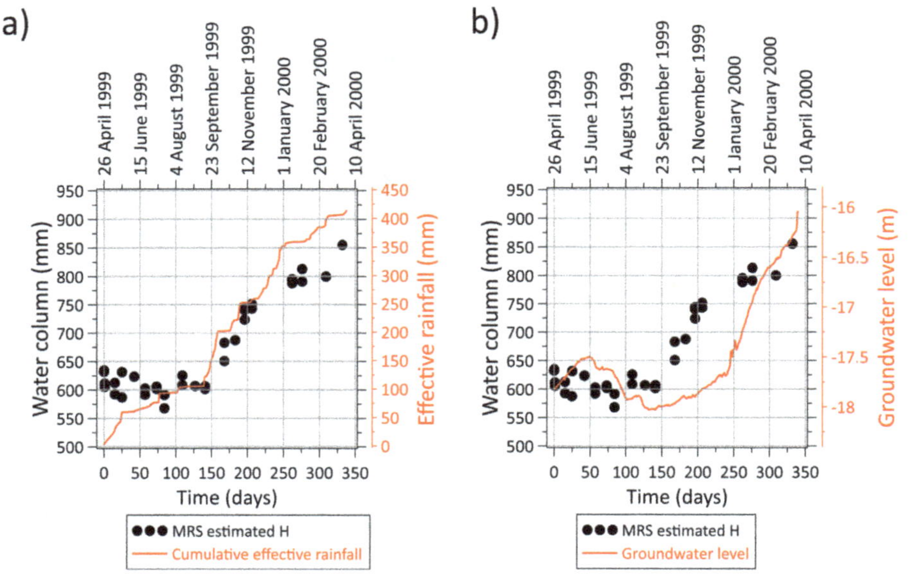


**Figure 7.31.** Villamblain test site: (a) the water column in the unsaturated zone estimated after the time-lapse MRS inversion versus monitoring time (black dots) and the effective rainfall provided by the meteorological observations (red line, red axis); (b) the MRS water column and the groundwater level monitoring (red line, red axis).

rainfall derived from the meteorological observations. We plot the cumulative effective rainfall using the red vertical axis. These graphs show that in summer months (May–August), the water column diminishes and hence water in the unsaturated zone contributes to the aquifers recharge. In September–December, the water column increases with the rate that follows the effective rainfall. At the beginning of the rain season, the rainwater does not flow into the aquifer, but recharges the unsaturated zone. Since January, the water column is about constant and the effective rainfall recharges the aquifer. During our study, the cumulative effective rainfall (413 mm) is shared between the increase of the water column in the unsaturated zone (211 mm) and the actual evapotranspiration plus the aquifer recharge (202 mm). We have no data for estimating the AET, and hence we cannot quantify the aquifer recharge. However, the mean annual recharge of the Beauce aquifer is about 110 mm and we consider our results in a reasonable agreement with the existing data. For comparison, the rise of GWL (figure 7.31(b)) shows a three-month delay relative to the effective rainfall, confirming that the MRS quantifies water in the unsaturated zone and not in the aquifer.

For modeling the one-dimensional vertical water flow, we use Richards' equation

$$\frac{\partial\theta(h)}{\partial t} = \frac{\partial}{\partial z}\left[K(h)\frac{\partial h}{\partial z} - K(h)\right] \tag{7.3}$$

where $h$ is the negative of the matric potential, $\theta$ is the volumetric water content, and $K$ is the unsaturated hydraulic conductivity. For solving Richards' equation, we use

HYDRUS-1 software and represent the soil water retention function $\theta(h)$ and the unsaturated hydraulic conductivity $K(h)$ in the functional form. For that, we use the Kosugi hydraulic property model. The log–normal distribution model allows for computing the water content as

$$S_e = \frac{\theta - \theta_r}{\theta_s - \theta_r} = \begin{cases} \frac{1}{2}\,\mathrm{erfc}\left[\frac{\ln(h/h_0)}{\sqrt{2}\sigma}\right] & (h < 0) \\ 1, & (h \geqslant 0) \end{cases} \tag{7.4}$$

where $S_e$ is the effective water content, erfc is the complementary error function, $h_0$ is the median metric head, $\sigma > 0$ denotes the standard deviation of the log-transformed soil pore radius and characterizes the width of the pore-size distribution.

The unsaturated hydraulic conductivity is

$$K(h) = \begin{cases} K_s S_e^l \left[\frac{1}{2}\,\mathrm{erfc}\left[\frac{\ln(h/h_0)}{\sqrt{2}\sigma} + \frac{\sigma}{\sqrt{2}}\right]\right]^2, & (h < 0) \\ K_s, & (h \geqslant 0) \end{cases} \tag{7.5}$$

where $K_s$ is the saturated hydraulic conductivity and $l$ is the pore connectivity parameter. Both the water content and the hydraulic conductivity are scale dependent parameters.

We calibrate the water flow model by minimizing the discrepancy between the theoretical and experimental MRS signals

$$\min_{\boldsymbol{\theta}} \left( \| \mathbf{A}\boldsymbol{\theta} - \mathbf{v} \|_{L2} \right) < \sigma_\varepsilon \tag{7.6}$$

where $\mathbf{v}$ is the measured amplitude of the MRS signal, $\boldsymbol{\theta}$ is the water content provided by the water flow model, and $\sigma_\varepsilon$ is an experimental error. We estimated the discrepancy between the theoretical and experimental data with the root-mean-square error

$$\mathrm{RMSE} = \sqrt{I^{-1}\sum_{i=1}^{I}\sum_{j=1}^{J}(a_{i,j}\theta_j - v_i)^2} \leqslant \sigma_\varepsilon.$$

Figure 7.32 shows the flowchart of the inverse modeling procedure.

The MRS data set comprised the measured amplitude of the MRS signal and the matrix $\mathbf{A}$. Field measurements also provided an estimate of the experimental error $\sigma_\varepsilon$. For computing the matrix $\mathbf{A}$, we considered the measuring conditions and the MRS measuring setup. The time-lapse MRS inversion for water content suggests the number and geometry of soil layers with different hydraulic parameters. These layers represent the subsurface in the hydraulic model. The hydrological data set comprises the water model parameters, the soil hydraulic property model, the soil hydraulic parameters, the upper and lower boundary conditions, and the variable boundary conditions. The first MRS sounding provides the initial water content profile. Knowledge of the geological

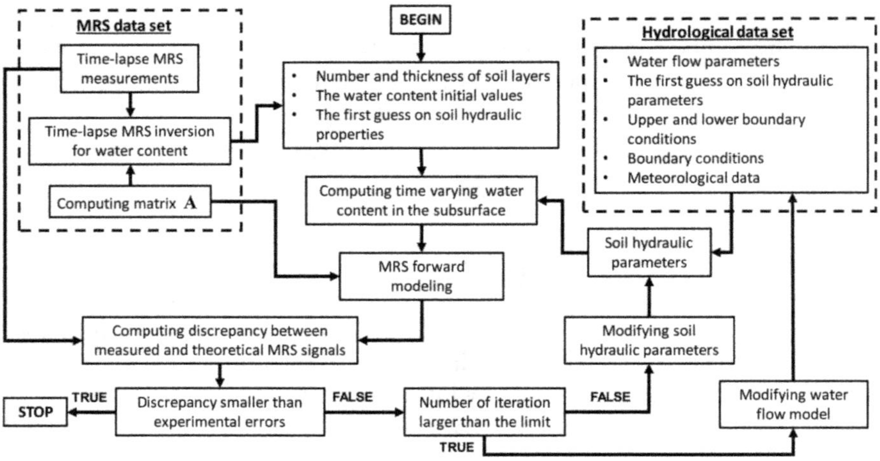


**Figure 7.32.** The flowchart of the unsaturated water flow modeling procedure considering time-lapse MRS measurements.

formations suggests the first guess for the soil hydraulic parameters. The meteorological data provide the daily precipitation and the potential evapotranspiration. The GWL depended on hydrological processes taking place at the watershed scale that is much larger compared to the scale of our study and we consider the GWL as an input variable in the hydraulic model. We iteratively adjust the soil hydraulic parameters, aiming to minimize the discrepancy between measured and theoretical MRS signals computed after the water content provided by the water flow model. For that, we apply the Levenberg–Marquardt algorithm. Iterations stop when the discrepancy becomes smaller than the experimental error. If the actual water flow model does not allow for solving equation (7.6), then we revise the entire hydraulic model.

In the water flow model, we use a simplified hydraulic model assuming a horizontal stratification. The MRS water content provided by the time-lapse inversion allows us to select four layers with different hydraulic parameters: at depths of 0–2 m, 2–12 m, 12–16 m and 16–20 m. The last layer below 16 m is almost saturated and, for this layer, the MRS provided the $\theta_s$. For the layers above the GWL, $\theta_s$ is a free parameter in inversion. In all four layers, the limestone is of the same geological origin and we assume, as the first guess, $\theta_s$ of these layers equal to that measured with MRS in the saturated limestone. Then, the inversion routine adjusted $\theta_s$. We have no data about the value of $\theta_r$. However, considering that MRS cannot see water with the water content close to $\theta_r$, we set $\theta_r = 0.001 \approx 0$ for all four layers. Note that MRS measures both $\theta_s$ and $\theta_r$ at a large scale and the results may be different compared to that measured on soil samples at a local scale. The MRS sounding at time zero provides the initial water content for the water flow modeling. Table 7.6 presents the soil hydraulic parameters in the water flow model that allows fitting the MRS measurements.

**Table 7.6.** The soil hydraulic parameters in Villamblain used in the unsaturated water flow model (the Kosugi hydraulic property model) that allows fitting time-lapse MRS data.

| Depth (m) | $\theta_r$ | $\theta_s$ | $\sigma$ (1/cm) | $n$ | $K_s$ (cm/day) | 1 |
|---|---|---|---|---|---|---|
| 0–2 | 0.001 | 0.1 | 500 | 1.6 | 15 | 0.5 |
| 2–12 | 0.001 | 0.15 | 250 | 0.8 | 3000 | 0.5 |
| 12–16 | 0.001 | 0.11 | 450 | 1.6 | 6 | 0.5 |
| 16–20 | 0.001 | 0.095 | 1000 | 2.2 | 1 | 0.5 |

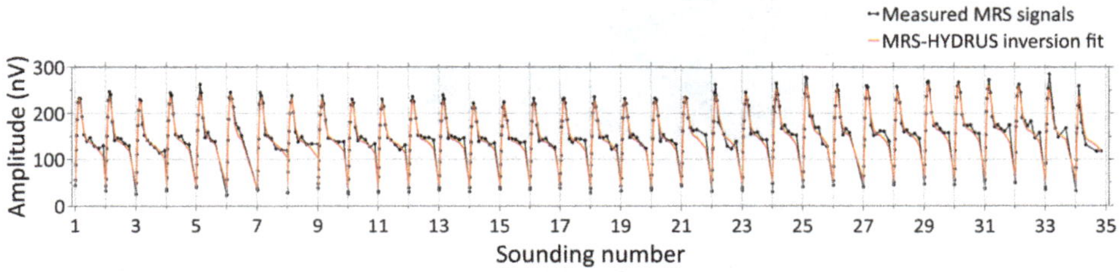


**Figure 7.33.** Measured amplitude of the MRS signal versus pulse moment for all soundings performed in Villamblain (black dots) and the theoretical amplitude computed after the water content provided by the unsaturated water flow model (red line). For each sounding, the pulse moment was normalized by the maximum pulse moment over all soundings.

Note that our hydraulic model aims to reproduce the water content observed with the MRS in Villamblain and not to assess the true hydraulic parameters of the subsurface. The layer between 2 and 12 m exhibits hydraulic parameters typical for a permeable material (coarse sand or fractured limestone). This layer has a high hydraulic conductivity and low water content when not saturated. The borehole points to an argillaceous limestone at this depth and we interpret this formation as fractured limestone containing some clay in the matrix. The MRS cannot see water in clay and shows no water in the limestone matrix. Fracturing rendered this formation permeable when saturated.

The theoretical amplitudes of the MRS signal computed after the water flow model fit measured amplitudes with the RMSE = 10.9 nV. The mean experimental noise is 14.2 nV. Figure 7.33 shows the amplitude versus pulse moment for all MRSs (black dots). The red line shows the amplitude computed using the water content provided by the unsaturated water flow model.

The time-lapse MRS inversion provides the water content (figure 7.30) consistent with that computed with the hydraulic model (figure 7.34). The root-mean-square error between the two images is 0.0079, which is close to the sensitivity threshold for MRS.

Figure 7.35 shows the water column computed between 0 and 18 m considering the water content provided by the time-lapse MRS inversion (black dots) and that given by the water flow model (red line).

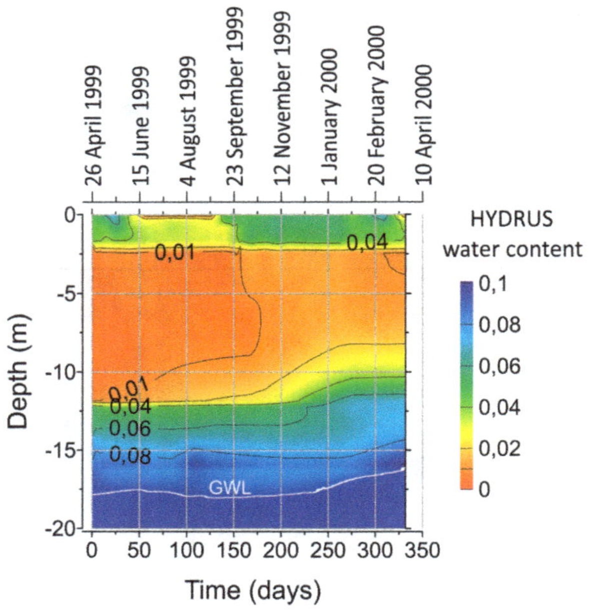


**Figure 7.34.** The water content given by the unsaturated water flow model computed with the HYDRUS-1D program.

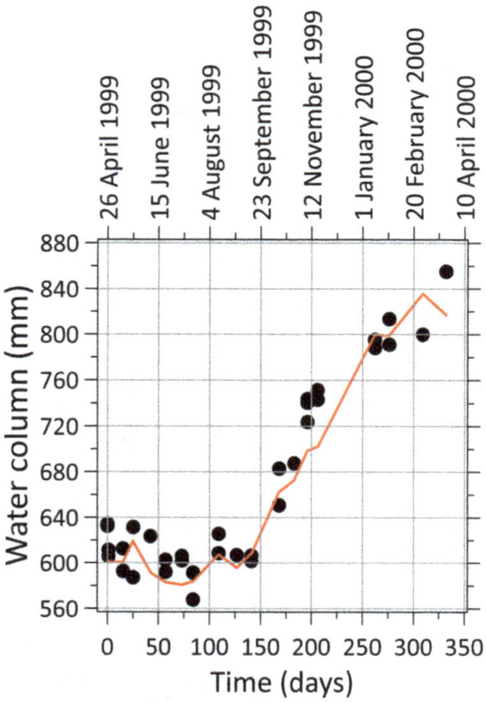


**Figure 7.35.** The water column in the depth interval between 0 and 18 m (the unsaturated zone) measured with MRS (circles) versus time. The red line shows the water column provided by the unsaturated water flow modeling.

Our results confirm the possibility of developing an unsaturated water flow model using time-lapse MRS measurements. Such a model allows for reproducing the water content profile observed with MRS. For making such a model representative in the investigated geological environment, one has to use more geological, hydrological and meteorological data.

CPSIA information can be obtained
at www.ICGtesting.com
Printed in the USA
BVHW052046250222
630126BV00003B/11